INNOVATIVE SYSTEMS for SEISMIC REPAIR & REHABILITATION of STRUCTURES

HOW TO ORDER THIS BOOK

BY PHONE: 800-233-9936 or 717-291-5609, 8AM–5PM Eastern Time

BY FAX: 717-295-4538

BY MAIL: Order Department
Technomic Publishing Company, Inc.
851 New Holland Avenue, Box 3535
Lancaster, PA 17604, U.S.A.

BY CREDIT CARD: American Express, VISA, MasterCard

BY WWW SITE: http://www.techpub.com

PERMISSION TO PHOTOCOPY—POLICY STATEMENT

Authorization to photocopy items for internal or personal use, or the internal or personal use of specific clients, is granted by Technomic Publishing Co., Inc. provided that the base fee of US $3.00 per copy, plus US $.25 per page is paid directly to Copyright Clearance Center, 222 Rosewood Drive, Danvers, MA 01923, USA. For those organizations that have been granted a photocopy license by CCC, a separate system of payment has been arranged. The fee code for users of the Transactional Reporting Service is 1-56676/00 $5.00 + $.25.

INNOVATIVE SYSTEMS for SEISMIC REPAIR & REHABILITATION of STRUCTURES
Design & Applications

Proceedings of
Second Conference on Seismic Repair
& Rehabilitation of Structures (SRRS2)

Fullerton, California, USA
March 21–22, 2000

EDITED BY
Ayman S. Mosallam, Ph.D., P.E.
Center of Seismic Repair & Rehabilitation of Structures (SRRS)
Division of Engineering
School of Engineering and Computer Science
Fullerton, California 92834
U.S.A.

Innovative Systems for Seismic Repair & Rehabilitation of Structures
a TECHNOMIC®publication

Technomic Publishing Company, Inc.
851 New Holland Avenue, Box 3535
Lancaster, Pennsylvania 17604 U.S.A.

Copyright © 2000 by California State University, Fullerton SRRS Center
All rights reserved

No part of this publication may be reproduced, stored in a
retrieval system, or transmitted, in any form or by any means,
electronic, mechanical, photocopying, recording, or otherwise,
without the prior written permission of the CSUF/SRRS Center.

Printed in the United States of America
10 9 8 7 6 5 4 3 2 1

Main entry under title:
  Innovative Systems for Seismic Repair & Rehabilitation of Structures:
    Design & Applications

A Technomic Publishing Company book
Bibliography: p.

Library of Congress Catalog Card No. 00-101387
ISBN No. 1-56676-964-7

Table of Contents

Preface ix

Advanced Composite Materials for 21st Century Bridges:
The Federal Highway Administration's Perspective 1
J. M. HOOKS

Applications of Polymer Composites in California
Highway Bridges 11
J. ROBERTS

Design Concepts for Composite Plate Bonding and
Column Confinement. 26
L. HOLLAWAY

Practical Implementation of Design Procedures for Retrofit of
Bridge Columns Using FRP 46
R. A. IMBSEN and F. ALAMEDDINE

Seismic Response of Reinforced Concrete Moment Connections
Repaired and Upgraded with FRP Composites 59
A. MOSALLAM, P. CHAKRABARTI, S. SIM
and H. M. ELSANADEDY

Seismic Retrofit of Reinforced Concrete Members with
CFRP Composites 73
M. A. ISSA, M. S. ISLAM, M. LESLIE, H. ABDALLA
and C. DO VALLE

Seismic Retrofit of Reinforced Concrete Columns Using FRP
Composite Laminates 85
M. HAROUN, M. FENG, M. YOUSSEF and A. MOSALLAM

The Retrofit Design of Concrete Columns and Slabs with Externally
Applied Fiber-Reinforced Polymer (FRP) Composite Materials 96
R. M. ELHASSAN

Prediction of Cyclic Performance of Composite-Jacketed
Squat Reinforced Concrete Bridge Columns 108
M. A. HAROUN and H. M. ELSANADEDY

Repair and Upgrade of R/C Two-Way Slab with
Carbon/Epoxy Laminates 119
A. MOSALLAM, T. LANCEY, J. KREINER, M. HAROUN
and H. ELSANADEDY

Strengthening of Full-Scale Reinforced Concrete Beams
Using FRP Laminates and Monitoring with Fiber Optic
Strain Gauges 131
D. D. McCURRY, JR. and D. KACHLAKEV

Seismic Repair and Rehabilitation Using FRP Composites:
A Systematic Approach 141
G. R. STEVENS

Design Philosophy for Strengthening with Carbon Fiber
Reinforced Polymer Composites 147
E. FRETT

Upgrade of Naval Wharf Bravo 25, Pearl Harbor, Hawaii 159
S. LANSBURG

Structural Upgrade and Repair of Wood Members Using
Cross-Ply Carbon/Epoxy 162
A. MOSALLAM, J. KREINER and T. LANCEY

Settlement Repair of Lightly Reinforced Concrete Block Walls
Using CFRP 171
G. MULLINS, A. HARTLEY, D. ENGEBRETSON and R. SEN

Composite Retrofit of Unreinforced CMU Walls 181
J. GERGELY, D. T. YOUNG, J. HOOKS and N. AL-CHAAR

A Crucial Link between Building Codes and Seismic
Strengthening Technologies. 187
B. N. HORECZKO

A Design Approach for FRP Composite Structural Shapes 197
V. SHEKAR, H. THIPPESWAMY and H. V. S. GANGARAO

**Repairing and Strengthening Reinforced Concrete Columns
Using Ferrocement Laminates** 210
E. H. FAHMY, Y. B. I. SHAHEEN and Y. S. KORANY

**Earthquake Rehabilitation of a Historic Concrete Structure
Using Fluid Viscous Dampers** 224
H. K. MIYAMOTO and D. A. LEE

**Seismic Connection Designs for Retrofitting
Steel Moment Frames**. 234
J. ALLEN and R. M. RICHARD

Author Index   239

Preface

In recent years, severe earthquakes throughout the globe have prompted the structural engineering community to pursue considerable research and development of innovative repair and rehabilitation systems. The recent Turkey and Taiwan earthquakes and the 7.0 "Non-damaging" earthquake of southern California served as grim reminders that while natural disasters cannot be prevented, engineers and researchers must devise methodologies for building structures that can withstand high magnitude quakes. Moreover, we must be able to retrofit existing buildings and infrastructures to minimize their vulnerability to seismic forces.

For the past few decades, several seismic repair and rehabilitation systems have been used including various epoxy injection, concrete and steel jacketing and plating, external steel prestressing, stitching, and addition of steel stiffeners. The primary function of these systems is to enhance the strength, stiffness, and the energy-dissipation characteristics of the reinforced concrete frame structures. However, there are several problems associated with the applications of these conventional repair and retrofitting systems. For example, the difficulty of bonding heavy weight steel plates, especially to horizontal surfaces, and the requirement of large spaces and heavy equipment for the applications of these systems. In addition, these systems will generally alter the architecture of the retrofitted facilities and will require longer application times.

Although the conference, in general, deals with state-of-the-art techniques for repair and rehabilitation of structures, special emphasis is given to the use of polymer composite systems as promising techniques for infrastructure repair and rehabilitation for the new millennium. The use of advanced materials in repair and rehabilitation of existing structures has been accepted by the construction industry. However, only a small sector of the industry has been exposed to such new technology. This conference serves as an avenue to disseminate information on the availability of these state-of-the-art technologies for seismic repair and rehabilitation of structures.

The objectives of this conference are (1) to expose the local and national construction industry to the ongoing research activities related to the development of state-of-the-art technologies for seismic repair and rehabilitation of structures, including buildings and bridges; (2) to reach out to the construction industry and create awareness among the key participants in the construction business (structural engineers, architects, construction contractors, insurance companies, etc.) about new methods for repair and retrofit of structures; (3) to provide the construction industry with the latest research and design information for using these systems; to update owners, architects, and structural engineers about different ongoing activities related to building codes, and standard specifications for the new systems including polymer composites; and (5) to provide an opportunity for direct contact among academia, industry, state, and government laboratories.

This proceedings contains twenty-two articles covering different topics related to seismic repair and rehabilitation of structures. The papers presented in the conference are subdivided into the following sessions: keynote papers, overview, seismic repair and rehabilitation of columns, seismic repair and rehabilitation of bridges, seismic repair and rehabilitation of columns, seismic repair and rehabilitation of beams and slabs, seismic repair and rehabilitation of columns, seismic repair and rehabilitation of connections, seismic repair and rehabilitation of masonry walls, non-composite repair and rehabilitation technologies, and building codes. The focus of all papers is on design philosophy and design methodology, including numerical design examples. The conference is sponsored by the *Structural Repair and Rehabilitation of Structures (SRRS) Center,* Division of Engineering of California State University at Fullerton and co-sponsored by thirty national and international organizations including Federal Highway Administration, Caltrans, ODOT, American Concrete Institute, American Society of Civil Engineers, Earthquake Engineering Research Institute, ICBO Evaluation Services and others. The support provided by these organizations is greatly appreciated. Appreciation is extended to all the authors for their valuable contributions—without those contributions the conference and these proceedings would not be possible. Special appreciation to Dr. R. Rocke, Dean of School of Engineering and Computer Science and Mr. Jim Roberts, the Honorary Chairman of the conference for their invaluable guidance and support. The contributions of Dr. Chandra Putcha, the Conference Co-Chairman, and the members of the Conference Organizing Committee, Dr. J. Kreiner, Chairman of the International Scientific Committee, Dr. T. Lancey, Division Chairman are greatly appreciated. The effort of Ms. K. Syre and the staff of ATD, and Ms. K. Chowbey for managing the conference, Ms. Doris Kalin of Comp Graphic for designing and maintaining the conference web page, and designing the conference brochures. The effort of Mr. J. Kiech of SRRS Center in preparing the initial conference flyer is highly appreciated.

Finally, special thanks go to my wife, Hanaa, for her patience and tremendous support and to my two children Tamer and Dean hoping that they will forgive me for being away from them preparing for this conference.

On behalf of the School of Engineering and Computer Science of California

State University, I hope that the conference has succeeded in meeting its objectives and that both the construction and the composite industries have benefitted from the information presented in this conference.

AYMAN S. MOSALLAM, PH.D., P.E.
Fullerton, California, USA
March, 2000

Advanced Composite Materials for 21st Century Bridges: The Federal Highway Administration's Perspective

J. M. HOOKS

ABSTRACT

More than 30 percent of the nation's 587,815 bridges are classified as deficient, i.e., deteriorated, under strength and/or geometrically obsolete for the demands of today's traffic volumes and loads. These deficient bridges and the repair work necessary to improve them represent a significant impediment to the nation's mobility and have a negative impact on productivity. The Federal Highway Administration (FHWA) is committed to the strategic goals of improving mobility and increasing productivity on the nation's highways. Regarding bridges, the performance measures for these goals are a significant reduction in the number of deficient bridges and a reduction in the time and cost necessary to complete new bridges and bridge improvement projects. Mobility will be improved by eliminating deficient bridges and reducing the "down time" necessary to complete a project. Productivity will be increased by eliminating detours and reducing the life cycle costs of building and maintaining bridges. One way to meet this challenge involves development and utilization of significantly improved bridge materials such as high performance steels and concretes. The bridge engineering community is also experimenting with a material new to the highway infrastructure - fiber-reinforced polymer (FRP) composites. FRP materials have been used extensively in the aerospace industry but only recently are they being applied on highway bridges. Among other properties, FRP materials have a high strength to weight ratio and excellent resistance to attack from chemicals including road salts. In addition, easy prefabrication of FRP bridge elements and comparatively short installation times, make FRP materials excellent candidates for bridge applications.

In June 1998, Congress passed the Transportation Efficiency Act for the 21st Century (TEA-21). This landmark legislation authorizes funds for two major initiatives intended to improve the condition, durability, and capacity of bridges. First, TEA-21 continues the Highway Bridge Rehabilitation and Replacement Program (HBRRP), which provides $20.4 billion to rehabilitate or replace bridges rated deficient. Second, a new initiative, the Innovative Bridge Research & Construction (IBRC) Program is launched. IBRC champions the use of innovative materials and technologies to repair, rehabilitate or replace bridges. In the first three years of the IBRC program, 116 projects utilizing innovative materials have been funded. Sixty-two of these projects involve the use of FRP composites in bridge decks, concrete reinforcement, prestressing tendons, column wrapping or bonded sheets for increased bridge strength.

Federal Highway Administration, 400 Seventh Street SW, Washington, D.C. 20590

THE NATION'S INVENTORY OF BRIDGES

Prior to the collapse of the Silver Bridge at Point Pleasant, West Virginia in December 1967, there was no comprehensive, nationwide database of information about the number, type, location and condition of the nation's bridges. The tragic loss of 47 lives at Point Pleasant soon led to legislation which mandated the National Bridge Inspection Standards and the creation of the National Bridge Inventory (NBI). The Federal Highway Administration (FHWA) is required to maintain a complete inventory of all highway bridges on public highways - a bridge is defined as a vehicular structure with a total span length of at least 20 feet along the centerline of the highway. One major function of the NBI is to maintain a complete inventory of bridge data related to location, type and geometry of each bridge; roadway or feature crossed, responsible owner, age, materials of construction, design load capacity, etc. The NBI makes it possible to search the database of the nation's bridges and extract data which help engineers and highway agency managers evaluate the status of bridges and develop programs to continually improve that status.

The Age of the Nation's Bridges

Slightly less than half (47.5%) of the bridges in the United States are owned and maintained by State Departments of Transportation (DOTs) including a small percentage owned by quasi-private roadway authorities. This includes all Interstate highway bridges and virtually all major structures such cable-stayed, suspension bridges, etcetera. The remaining 52.5% are owned and maintained by local highway agencies in the cities and counties. These latter bridges tend to be shorter and narrower than average, older, and often designed for live loads that are lower than the current standard of HS-25.

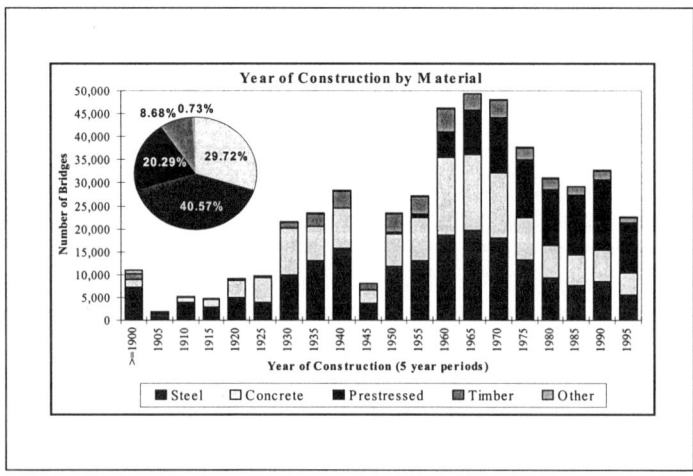

FIGURE 1. Age Range of Existing Bridges by Primary Construction Material

The range of ages of existing bridges is very wide, from bridges built within the last year to bridges that have been in service for more than 100 years. Figure 1 is a histogram of numbers of existing bridges that were built in each five-year period in the last 100 years plus all bridges built prior to 1900. The age of a bridge is not a definitive indication of current condition or capacity of a bridge - other factors such as environment, traffic volumes, maintenance history play a strong role. However, age is still a useful indicator when examining a bridge population for potential deficiencies.

The Condition of the Nation's Bridges

The second major purpose for the NBI is to maintain bridge inspection data which describes the current condition and capacity of the major elements of every bridge. This data is virtually up to date since all bridges on public highways must be inspected at least every two years by qualified inspectors. Inventory information is verified or updated and the inspectors rate the condition of the bridge elements on a scale from zero (closed) through nine (new). Each state DOT sends an update to the FHWA every year and the new data is used to make the NBI records current. The information also includes estimated costs of rehabilitation or replacement as necessary. FHWA is able to make a uniform assessment of the status of bridges on a state-by-state basis as well as for the nation as a whole.

By law, the data in the NBI is used to apportion federal Highway Bridge Rehabilitation and Replacement Program (HBRRP) funds to each state on the basis of the total needs for bridge improvement projects. However, the eligibility of any specific bridge project for funding depends on an assessment of various factors. The so-called Sufficiency Formula is used to calculate a numerical sufficiency rating for each bridge in the NBI.

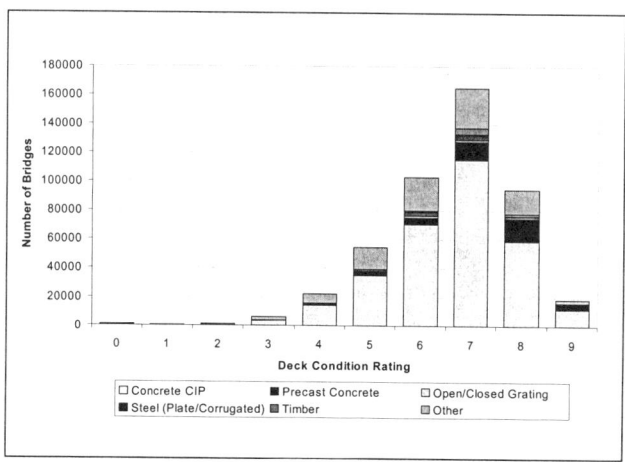

FIGURE 2. Deck Condition Rating Histogram

Bridges with a sufficiency rating of less than 80 are eligible for HBRRP funds for rehabilitation; a bridge is eligible for replacement with HBRRP funds if the sufficiency rating is less than 50. The information in the NBI can be examined to provide insight into what types of deficiencies are common. Figure 2 provides data on condition rating versus age of existing bridge decks, by material. When combined with available traffic data, the need for deck rehabilitation projects where minimal bridge closures are critical can be determined. These results are quite useful in assessing the potential role of FRP composites in future repair and rehabilitation of bridges. For instance, the "market" for precast, quickly installed precast deck panels can be assessed. Further collation with load capacity data can pinpoint the added need to decrease dead weight in order to raise live load capacity without major structural strengthening.

FHWA'S STRATEGIC GOALS

The FHWA has five strategic goals and our progress in achieving them will contribute to achieving the overall transportation goals of the Nation. Two of these goals are defined as: mobility - continually improve the public's access to activities, goods, and services through preservation, improvement, and expansion of the highway transportation system and enhancement of its operations, efficiency, and intermodal connections; and productivity - continuously improve the economic efficiency of the Nation's transportation system to enhance America's position in the global economy. Some of FHWA's goals are interrelated, e.g., improving the infrastructure and operations of the highway system promotes productivity, safety, and national security as well as mobility.

Performance indicators have been established to help measure FHWA's progress toward achieving those goals. As related to highway bridges, those performance indicators include: preserving and enhancing the infrastructure of Federal-aid highways and bridges by improving the condition of all bridges so that less than 25 percent are classified as deficient in 10 years; reducing delays on Federal-aid highways by 20 percent in 10 years including delays caused by bridge repair and rehabilitation projects; minimizing the time needed to return bridges to full service following disasters; improving the return on investment of the highway system by reducing the life cycle costs of building and maintaining bridges.

Improving Mobility and Productivity

In order to reach those goals of improved mobility and productivity, FHWA strategies will focus on: enhancing the infrastructure by focusing research and technology innovations on ways to make transportation investments buy more and last longer. In order to meet our objectives for bridge improvements, we will leverage research to foster major advances in the technology of bridge construction, repair, and maintenance. FHWA technology deployment initiatives will ensure that current advancements such as high performance materials (including composites) are adopted to improve the performance of bridges. We work with our Federal, State, and local partners to ensure that highways can provide vital links for emergency relief during natural disasters and to ensure that full highway access is quickly restored to the disaster area.

Minimizing the cost to build, maintain, operate, and use the National Highway System directly supports local, regional, and national economic growth and competitiveness. The Highway System provides the majority of passenger travel; 91 percent of the person-distance (miles) traveled are in privately owned vehicles. Trucks move nearly three quarters of the value, one half of the weight, and nearly one quarter of the metric ton- kilometer (ton-miles) of all freight shipments in the United States and its territories. FHWA programs to increase mobility through enhanced infrastructure, technology, and operations also support economic performance goals. In addition, we will help reduce the economic costs of providing and using the highways by increasing the return on investment of highway dollars through research and technology transfer to significantly reduce the life cycle costs of new and reconstructed bridges.

Future Demands

FHWA's strategic goals and objectives were developed based on assumptions about future demands on the highway system. The forecasted trends in personal transport and commercial transport show particularly significant growth in vehicle-distance traveled. Demographic studies indicate that as the population increases, travel demands will increase and change. The Bureau of the Census estimates a 21-percent increase in population by 2020 and a 47-percent increase by 2050. The increase in population will increase the number of trips and distance traveled. Between the 1970 and 1990 census, 91 percent of the total U.S. population growth has been in the suburbs and a large percentage of the new jobs created were also in the suburbs. The analysis supporting the FHWA's Condition and Performance Report projects that over the next 10 years the vehicle distance (miles) traveled is estimated to increase by 24 percent and by 53 percent in 20 years. More people, more cars, more miles traveled will add more strain to all aspects of the highway system. As for commercial transport, a significant portion of commercial freight is moved on the highways. The number of commercial trucks on America's highways grew by 76 percent between 1982 and 1992, while vehicle distance traveled doubled. To be competitive in the global economy, U.S. producers must maximize the efficiency of production and distribution. For example, just-in-time delivery systems have greatly reduced overhead costs and freight logistics systems have increased efficiency. As manufacturers rely more extensively on improved logistics to increase economic efficiency, demands on highway capacity and reliability increase. All of these facts will continue to produce pressure to provide bridges which are less costly and more maintenance free.

The Role of FRP Materials in Bridges

Traditional bridge building materials, primarily steel, reinforced concrete and prestressed concrete, are well understood by bridge engineers. Design codes are well established and experience has proven that bridges built of the materials can serve for many years. Additionally, new advancements in strength and durability of these materials has enhanced their role in bridge building. However, it is still clear that even if FRP composite materials are not as well understood as steel and concrete, they have several superior characteristics which are highly desirable in bridge applications.

These characteristics include: high strength combined with light weight; high resistance to fatigue; high resistance to corrosion from deicing salts as well as other chemical agents; ease of prefabrication and ease of shipping and jobsite handling of elements and modules; ease and speed of installation. All of these characteristics can be put to good use in bridge rehabilitation and in bridge construction with positive impacts on mobility and productivity. For example, the modular panel construction of FRP bridge deck systems enable quick project delivery. A bridge deck fabricated of composite materials can be constructed and put in service in a relatively short time and at a competitive cost. The lightweight material and ease of construction provide tremendous labor and traffic control cost savings. This technology has demonstrated that a bridge structure can be replaced and put into service in a matter of days rather than weeks or months. In many situations, a replacement deck of FRP materials could reduce the weight of conventional construction by 70 to 80 percent. This reduction in dead weight can be used to allow an increased live load, thereby eliminating a previously load posted structure without major modification to the superstructure. In other situations, understrength bridge members can be upgraded using bonded sheets of FRP laminates and load limits may be lifted.

THE TRANSPORTATION EQUITY ACT FOR THE 21ST CENTURY

Transportation legislation is authorized on a six-year basis. From fiscal year 1991 through fiscal year 1997, the Intermodal Surface Transportation Equity Act (ISTEA) was the governing legislation for transportation. Under ISTEA, several experimental applications of innovative materials, including FRP, were constructed on bridge projects. Transportation legislation was reauthorized in June 1998 when Congress passed and President Clinton signed the Transportation Equity Act for the 21st Century (TEA-21). This landmark legislation authorizes up to 162 billion dollars for highway and bridge construction over six years. Included in the reauthorization was the continuation of a major initiative intended to improve the condition, durability, and capacity of the Nation's 589,815 bridges. Since 1979, the Highway Bridge Rehabilitation and Replacement Program (HBRRP) has been instrumental in improving or replacing more than 52,000 bridges. TEA-21 authorizes $20.4 billion to rehabilitate or replace bridges judged to be eligible because of deteriorated condition and/or reduced capacity. Certainly, the effectiveness of the investment of the past HBRRP funds cannot be simply judged on the sole basis of the number of deficient bridges which were upgraded. Equally important as the numbers are the efficiency with which the funds were used (productivity), the minimization of construction or rehabilitation times (mobility and productivity) and the ultimate extension of service life of the repaired or new bridges (mobility and productivity). One of FHWA's concerns is that the HBRRP funds are used to repair bridges or build new bridges which will have a longer, maintenance free service life.

The Innovative Bridge Research & Construction (IBRC) Program

The Transportation Equity Act for the 21st Century (TEA-21) launched an important new initiative in the effort to reduce the number of bridges in the United States which are rated as deficient, i.e., functionally obsolete and/or structurally deficient. TEA-21 establishes the Innovative Bridge Research and Construction (IBRC) Program which provides $102 million over six years for bridge projects which demonstrate the application of innovative materials such as FRP composites. As stated in the legislation, under the program, the Secretary (of Transportation) shall make grants to, and enter into cooperative agreements and contracts with - - - States to pay the Federal share of the cost of repair, rehabilitation, replacement, and new construction of bridges or structures that demonstrate the application of innovative materials. The Secretary shall select and approve the applications based on whether the project that is the subject of the grant meets the goals of the program, described below.

The goals of the IBRC program, as established by the Congress, are consistent with the strategic goals of the Department of Transportation which are to increase mobility in the transportation system and improve productivity - one of the key performance measures for assessing progress on these goals is the reduction of the number of bridges which are rated as deficient. In part, the Congress described the IBRC program goals as including (A) the development of new, cost-effective bridge applications of innovative materials; (B) the reduction of maintenance costs and life-cycle costs of bridges, including the costs of new construction, replacement, or rehabilitation of deficient bridges; (C) the development of construction techniques to increase safety and reduce construction time and traffic congestion; (D) the development of engineering design criteria for innovative products and materials for use in highway bridges and structures; and (E) the development of highway bridges and structures that will withstand natural disasters, including alternative processes for the seismic retrofit of bridges. Section 5103 of TEA-21 authorized funds to be available to the States for projects to demonstrate innovative materials relating to repair, rehabilitation, and construction of bridges. Eligibility is governed by the following:

- Funds are available for bridge projects that meet one or more of the seven program goals listed in Section 503(b)(2) of Title V- Transportation Research

- The project may be on any public roadway, including State and locally funded projects.

- Funds are available for costs of preliminary engineering, costs of repair, rehabilitation or construction of bridges or other structures, and costs of project performance evaluations including instrumentation and performance monitoring of the structure following construction.

- Proprietary Products - As this is a research and experimental program, it is in the public interest that proprietary and sole source products may be included in the projects, but they must be clearly identified and described.

These funds may be used for the Federal share of the cost of the repairs, rehabilitation, replacement or new construction on the "innovative materials" portion of the project.

The program was initiated by the Federal Highway Administration (FHWA) in July 1998 by solicitation to all of the States to identify and submit for funding, projects which utilize innovative materials and which meet one or more of the program goals. A subsequent call for candidate projects was made in May 1999. A total of 41 States responded and a total of 289 projects were submitted; the combined total estimated cost of the projects was $172M. The selection criteria for eligible projects required the application of a material considered innovative in highway bridges and were also keyed to how well the application met the program goals.

A closer look at the types of candidate projects indicates that FRP composite materials are clearly capturing the imagination and engaging the engineering capabilities of the nation's bridge engineers. As a result of the two calls for candidate projects, 165 of the 289 candidate projects involved an application of FRP composites in either new, replacement or rehabilitated bridge components. Figure 3 indicates how many projects include each of the six main types of applications identified.

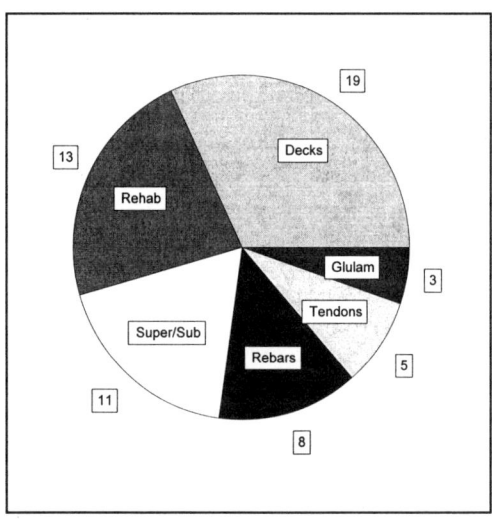

FIGURE 3. Distribution of FRP Composite Applications in IBRC Program, 1998 to 2000

Preliminary results indicate that almost two thirds of the projects selected for funding under the first solicitation will involve applications of FRP composites. Types of projects being submitted include: full depth deck sections for new structures as well as replacement decks for deficient bridges; column wrapping for deteriorated and/or seismically vulnerable bridges; composite reinforcing bars for concrete elements subject to corrosion of the reinforcing steel; and repairs and strengthening of concrete slabs and beams.

SOME IBRC PROJECTS WITH APPLICATIONS OF FRP COMPOSITES

Salem Avenue Bridge in Dayton, Ohio

One well-recognized model of incorporating new materials and technologies in bridge construction is the Salem Avenue Bridge (Montgomery SR 49) in Dayton Ohio. When completed, this twin 5-span, 680-foot-long, 48-foot-wide bridge will incorporate the largest composite bridge deck in the United States—65,280 square feet. In addition to the Ohio Department of Transportation (ODOT), partners include the FHWA, HITEC, The Composites Institute, Montgomery County Engineers, four universities, the U.S. Army Corps of Engineers, and four manufacturers. For example, environmental testing and materials testing will be by the University of Maine. The University of Kentucky tested the deck panels, and the University of Cincinnati and the Ohio University were responsible for field monitoring. The Corps of Engineers is conducting 10 million cycles of fatigue testing on the bridge materials. Half of the materials will be tested at -22 °F, the other half will be tested at 120 °F. Fiber optics will provide long-term monitoring of the internal structure.

Alabama State Highway 81 over Uphapee Creek.

This three-span (48 feet-65 feet - 48 feet) reinforced concrete bridge was built in 1945. The reinforced concrete girders in the continuous spans are exhibiting advanced symptoms of distress caused by routine heavy truck traffic over the life of the bridge. The bridge was rated in 1994, resulting in a posted load limit for the standard tri-axle load of 25.4 tons, 12 tons less than the standard load. The proposed application will be to rehabilitate the girders by externally bonding FRP laminates to the bottoms and sides of the girders. It is anticipated that the rehabilitation will allow the load limits to be lifted at a small fraction of time and cost of other improvement schemes.

Maryland 24 over Deer Creek.

This simple span (123 feet long) steel truss bridge is similar to many such bridges in Maryland, some of which are designated as historical structures. Core samples of the reinforced concrete deck indicate water seeping through the deck. More than 55% of the deck is patched and the steel truss has large sections which are rusted. The proposed FRP application will be a full deck replacement with deck sections prefabricated from FRP composite materials. The proposed application will provide a new, durable and lightweight deck which should be maintenance free for many years. In similar situations, the lightweight deck may also allow existing load limits to be lifted without disturbing the historic nature of the truss bridge.

Old Youngs Bay Bridge, Oregon

This double leaf bascule bridge with a total length on the draw spans of 150 feet was built in 1921. The draw spans are steel trusses with a mass concrete counterweight. The existing deck on the lift spans is laminated timber. The proposed FRP application will replace the existing timber deck with a pultruded modular FRP composite deck. The FRP deck is a desirable solution because its weight closely approximates the weight of the existing deck, thereby eliminating the need to significantly modify the counterweights on the draw spans.

SUMMARY

The IBRC program will continue for the duration of the TEA-21 legislation and three additional solicitations are planned beginning on March 15, 2000 and continuing on each March 15th through the year 2002. Program funds available for future projects amount to $60M. FHWA will continue to administer the program and will continue to stress that each project should include a carefully planned program of monitoring and evaluating the performance of the innovative materials. This will be critically important to the future use of FRP composites, since the knowledge base of their performance in bridge applications is very limited.

To learn more about the Innovative Bridge Research and Construction Program, including the solicitation and selection of projects for FY 2000 and beyond, visit http://www.ibrc.fhwa.dot.gov, and click on IBRC. This web page will provide detailed information on projects under way, as well as related research, development, and technology transfer efforts to enhance the use of innovative materials.

Applications of Polymer Composites in California Highway Bridges

J. ROBERTS

ABSTRACT

The California Department of Transportation has been engaged in cooperative research with the University of California at San Diego for the past six years to develop field applications of advanced composite materials for both repair of older structures and construction of new bridges.

The most highly developed application to date is the use of advanced composites in repair of bridge columns and other supporting elements to improve their ductility for seismic resistance. Both epoxy impregnated fiberglass and carbon fiber materials have been tested in the laboratory on half-scale models of bridge columns to determine the ductility that can be achieved in an older, non-ductile concrete column. The tests have confirmed the viability of these materials for strengthening existing structures and field application quality specifications have been developed. Since March, 1996 these specifications have been published and included as alternatives in over 50% of the seismic retrofit strengthening contracts advertised for construction.

The more exciting application of advanced composites is for new bridges and bridge deck replacement units. The research conducted so far has resulted in the design of a highway bridge composed of three foot diameter carbon fiber tubular bridge girders and a fully advanced composite bridge deck. Development of these elements has been underway for three years and laboratory testing is currently underway. The bridge design will be utilized on two state highway bridges in Southern California, to be advertised for construction in November, 1996. Further development of bridge deck replacement elements composed of advanced composite materials is continuing, with emphasis now on the connection details.

Although these advanced composite materials are expensive, the long life expected and their resistance to corrosion makes them competitive if the life cycle cost of a bridge in a highly corrosive environment is considered.

Future plans in the Caltrans-UC San Diego-ARPA-FHWA cooperative research program include the construction of a fully composite vehicle bridge on the UCSD campus which will cross over Interstate 5 north of San Diego. Construction of the smaller bridges is a preliminary step in the development and testing of the various components which will be utilized on this larger bridge.

SUMMARY

The California Department of Transportation has funded research at the University of California at San Diego for the past six years to develop field applications of advanced composite materials for both repair of older structures and construction of new bridges. The most highly developed application to date is the use of advanced composites in repair of bridge columns and other supporting elements to improve their ductility for seismic resistance. Epoxy impregnated fiberglass and carbon fiber materials have been tested in the laboratory on half-scale models of bridge columns to determine the ductility that can be

---

Chief Deputy Director and Chief Structures Engineer, California Department of Transportation, Sacramento, CA 94274-0001

achieved in an older, non-ductile concrete column. The tests have confirmed the viability of these materials for strengthening existing structures and field application quality control specifications have been developed.

INTRODUCTION

Transportation (Caltrans) began a research program, in cooperation with the University of California at San Diego (UCSD), to develop techniques for utilizing epoxy impregnated fiberglass sheets to wrap around older, non-ductile concrete bridge columns as an alternative to the already proven steel jacket technique. The jackets provide sufficient confinement in the concrete to allow them to perform in a ductile manner under seismic loading. It was known that the Japanese had used high strength carbon strands to similarly reinforce industrial stacks and chimneys but the use of glass fiber sheeting had not been used. The major unknown was the durability of the fiberglass materials under cyclic loading and to what level of ductility the columns could be designed. The testing program was conducted under the same conditions that were used in the testing of steel plate jackets. Half scale models of the prototype bridge columns were constructed, wrapped with the desired layers of glass fiber sheets and tested through several cycles of loading at various levels of ductility until the column failed due to degradation of its hysteretic performance. These laboratory tests proved that the epoxy impregnated fiber glass column Following the October, 1989 Loma Prieta earthquake the California Department of wraps could develop nearly the same ductile performance as the steel plate jackets.

Material properties are readily available from the manufacturers but there remained the issue of adequate quality control specifications for the field application. These early applications were rather crude, being hand laid in a similar manner as hanging wallpaper. It required some months to fully develop adequate quality control (QC) specifications so the materials tested in the laboratory could be replicated with confidence in the field. The application using epoxy impregnated fiber glass has been approved for two systems and field applications have been in place for over five years.

In 1993, following the end of the cold war and reduction of major aerospace and defense applications, the advanced composites industry began looking for applications of advanced composites in the civil infrastructure. The Caltrans-UCSD testing program was expanded to develop similar applications for the higher strength carbon fibers. This testing program has continued as more manufacturers submit their materials for approval and there are at least five systems approved for field application in California at this time. The carbon fibers are applied by automatic wrapping machines which wrap several 1/4 inch strands simultaneously and can fully wrap a typical four to six foot diameter, 20 foot long bridge column in two hours. Because of the higher strength to weight ratio these materials are very competitive with the steel shell retrofit technique, and they can be applied with much less heavy lifting equipment. The materials are much more resistant to corrosion than the steel jackets and they will require very little maintenance.

Working in cooperation with the University of California at San Diego research team and the ARPA and FHWA technology transfer programs we have been testing other applications of advanced composites in the seismic reinforcing of older bridges and in the construction of major bridge components and ultimately, a complete highway bridge designed for AASHTO loads. The first applications involve resin impregnated fiberglass or carbon sheets on non circular bridge members. These include the use of sheets to wrap and confine the spandrel columns and rib members on several arch bridges where it is difficult to access the locations with heavy equipment. The second application involves the use of small diameter carbon fiber tubes, constructed by the same technology as rocket bodies, for bridge girders. This application has been tested at the laboratory and design details are being developed for a bridge on the state highway system in southern California. The bridge will include deck units which are composed entirely of advanced composite materials and construction is scheduled for 2000. The testing program for these bridge components has been underway at UC San Diego for over three years, under the ARPA grant.

COLUMN STRENGTHENING

The most widely used application of advanced composite materials for bridges in California and other states, to date, is the seismic strengthening of bridge columns to improve their ductile performance in an earthquake. However, there is a larger market for this technology in the simple repair and strengthening of columns which have deteriorated from corrosion. It is relatively easy to clean and repair these columns and encase them with the non-corrosive composite materials. This application will undoubtedly increase the life of the columns or piers. Three manufacturers have developed prefabricated resin impregnated-fiberglass shells which can also be used as the form for concrete in the repair process. Figure 1 illustrates the prefabricated fiberglass column shell which has been approved and can be used also for repair of piling below the water line.

Figure 1 Installing Prefabricated Fiberglass Shell

Figure 2b) shows the clamping system utilized to hold the pre-fabricated shell tight until the adhesive cures. This system is the "Clockspring" system, utilizing an Isothalic Polyester resin. Several layers of shells are applied to provide the required ductility. Figure 2a) shows the installation of a full height prefabricated shell. This is the "Du-Pont Hardcore" system, utilizing a Vinylester resin. These applications were installed in 1996 on the Santa Monica Freeway, Interstate 10, in Los Angeles.

A third prefabricated system has been developed by NCF which utilizes several layers of four foot high single shells. The system, known as "Snaptite" is fabricated and heat cured on a mandrel under controlled curing conditions in the manufacturing plant and shipped to the field much the same as the two systems illustrated in figures 1and 2. This system appears much less cumbersome to install than the other two prefabricated systems.

Figure 2 b) Installing Full Height Shell   *Figure 2a) Clamping the Pre-Fabricated Shell In Place*

Figure 3 shows the use of epoxy-fiberglass as a confinement membrane to increase column ductility and toughness. This was the first application to be tested and approved in California. The material has been used for both circular and rectangular columns. The aspect ratio of the rectangular columns cannot be more than 2:1 or the longer face will buckle under dynamic loading and the needed confinement will not be maintained. This material can be applied as a pre-preg or dry application with the epoxy being applied in the field. One of the initial problems with these materials was uniformity of the final appearance because they are hand laid sheets about three feet wide. Final appearance is dependent on the expertise of the field crew. Attempts have been made to design a machine to improve the application and insure more uniformity, but we have not seen that machine in use in California yet. Careful quality control of the field application and material mixing is necessary to guarantee a quality final product. Figure 4 shows the same material after field application and painting have been completed. The paint serves two functions; one is protection from ultra-violet light and the second is for aesthetics. The concrete colored paint does an excellent job for both functions. This application was implemented in 1991 on the Glendale Freeway (State Route 134) in Los Angeles. These materials had been tested at the UC San Diego Powell Laboratories in 1990 with excellent results. Both shear and moment ductilities of over eight (8) have been achieved in these tests.

Figure 5 illustrates the application of pre-preg carbon fiber wrapping on bridge columns at the field test site. The wrapping machine does not require heavy lifting equipment and a later version now applies more strands simultaneously but can wrap a column of 20 foot height in two hours. The columns are heat cured under controlled conditions by electrically heated blankets or enclosures. The columns are painted concrete color for aesthetic purposes, but the coating does provide protection against the elements. This system has been developed by XXSys Technologies and a second, similar system is being tested by Mitsubishi Industries. The thickness can be varied as the ductility requirements dictate. In the field applications on the Santa Monica Freeway the white paint was also used for the same purposes as on the resin impregnated fiberglass wraps. Figure 6 shows the field application on a seismic retrofit project in San Diego. This material has the potential of becoming the most cost effective column wrapping system because of its high strength to weight ratio. The system does not require heavy lifting equipment and can generally compete favorably against the steel shell retrofit systems. Since it is not as labor intensive as some of the other systems being approved, it will ultimately be the system of choice for most contractors. The controlled heat curing system that is used by XXSys provides a material that is very reliable and has the best chance of guaranteeing the same properties as those of the laboratory samples. This relaiability is more difficult to achieve with many of the other systems.

Figure 3 Epoxy-Fiberglass Column Wrap in Laboratory Test

Figure 4 Field Application

Figure 5 Application of Carbon Fiber Column Wrap

Figure 6 Completed Carbon Fiber Wrap

STRENGTHENING ARCH RIBS AND SPANDREL COLUMNS

Analysis has shown that the typical arch bridge is extremely vulnerable to seismic forces in the transverse direction and, therefore, it is necessary to retrofit strengthen the main ribs and spandrel columns. Getting any type of erection and application equipment into these locations is very costly and difficult so Caltrans has developed retrofit solutions using carbon or fiberglass sheets. Portions of the arch ribs and the joint where spandrel columns frame into the ribs are being reinforced by wrapping with these sheets. Testing has also been conducted at UC San Diego to evaluate the various sheet systems over the past three years. Results show that the shear capacity increase and confinement can be achieved to the levels required. Full scale tests were conducted during (1997) on the column-rib joint to determine the material thickness needed to provide the required performance in a seismic event. Use of these materials on older arches is also important from a historical preservation perspective because the sheets do not significantly alter the appearance of the bridges.

HIGHWAY BRIDGES

An innovative highway bridge consisting of hollow carbon-composite tubes and lightweight concrete will be built on State Highway Route 86 near Palm Springs, California with the construction to begin in 2000. The 64 foot long demonstration project was designed cooperatively by Doctor Freider Seible, Professor of Structural Engineering at UC San Diego and California Department of Transportation Bridge Design Engineers. The department will also monitor the bridge for long term performance in comparison with an existing parallel concrete bridge. The project is part of a $9.1 million Defense Advanced Research Projects Agency (DARPA) grant to UCSD and is supported by the Federal Highway Administration through its Technology Applications Office. The design is based on Dr. Seible's easy-to-assemble modular system of hollow tubes, each created of wound carbon fibers and filled with lightweight concrete. The girders and columns will be lightweight concrete filled to facilitate the deck to girder connection details. The deck will be made of a combination of concrete and advanced carbon fiber composites.

Figure 7 is a layout and plan of the bridge showing the basic components and dimensions. The bridge will carry normal AASHTO loads and is on a route that carries heavy truck traffic from the agricultural fields in the Coachella Valley. Figure 8 shows the basic principles of the carbon fiber tubes and concrete infill. The tubes are manufactured by the same methods that are used in manufacturing rocket body shells for the aerospace industry. Both machine wrapped and hand laid wrap tubes have been tested at the laboratory at UC San Diego. Figure 9 is a blowup of the deck and component connection details. The deck units have been undergoing development and testing at the laboratory for the past two years. The major work that remains is proof testing connection details.

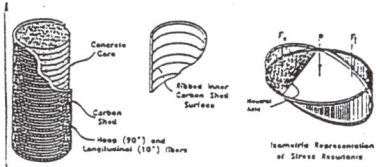

Figure 8 Carbon Tube Details

The proposed connection method is to drill holes through the deck units at the dowel locations and fill the space with in-situ polymer concrete. The safety railing for the initial bridge will be made of normal weight concrete but it is not unreasonable to assume that the railing could also be made of advanced composite materials.

Figure 10 shows the various types of deck panels that have already been tested. The final design configuration will probably be determined by economics and constructibility as much as by structural requirements.

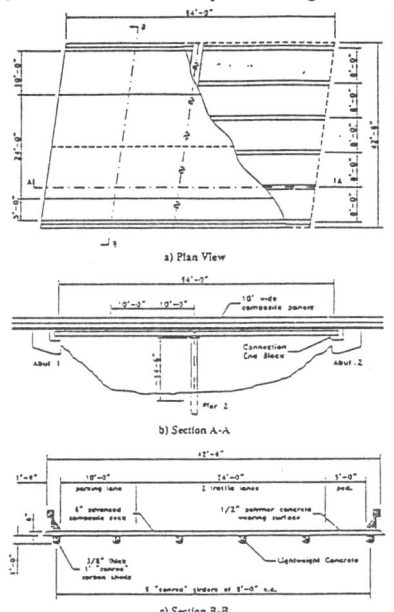

Figure 7 Layout of Bridge

The researchers at UC San Diego have installed four of the deck panels side by side on a campus access road at the laboratory to test their durability under normal traffic, including heavy trucks. The clear span on the test bridge is 15 feet.

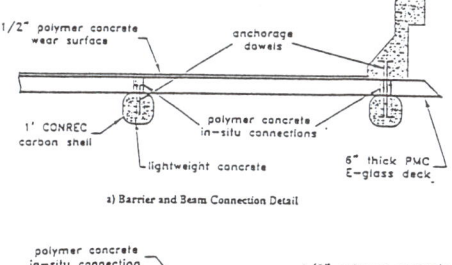

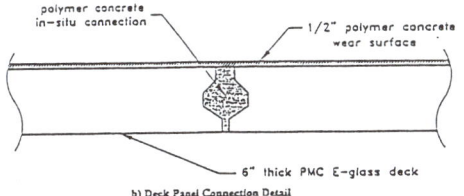

Figure 9 Connection Details

Figure 11 shows the carbon-concrete tube in the test stand during the flexural testing. Figures 12 and 13 are details of the connections at the bents & abutments. The most important need is the development of reliable connection details for the dissimilar materials. This bridge will be heavily instrumented and will provide much valuable information on the actual field performance of these details and the composite materials.

Figure (10): Composite Deck Panel Units

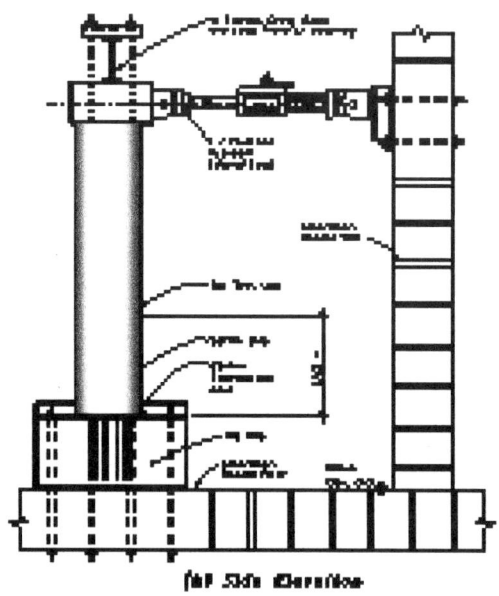

Figure (11): Carbon-Concrete Tube in Test Stand

a) Longitudinal Capbeam Connection Detail

b) Transverse Capbeam Connection Detail

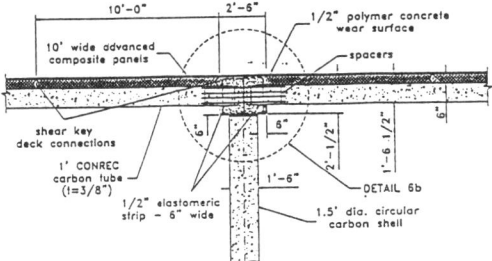

Figure 12 Bent Cap Details

a) End Block Anchorage Detail

b) Cap-Beam Joint Detail

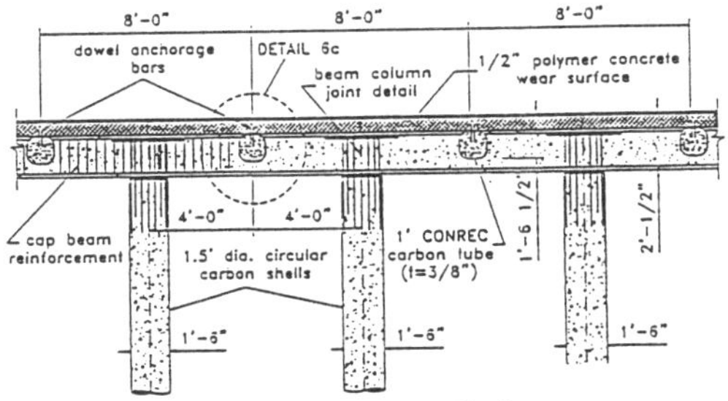

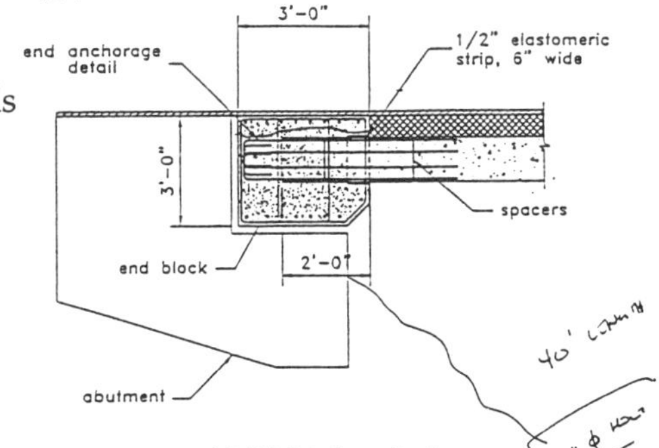

Figure 13 Joint Details

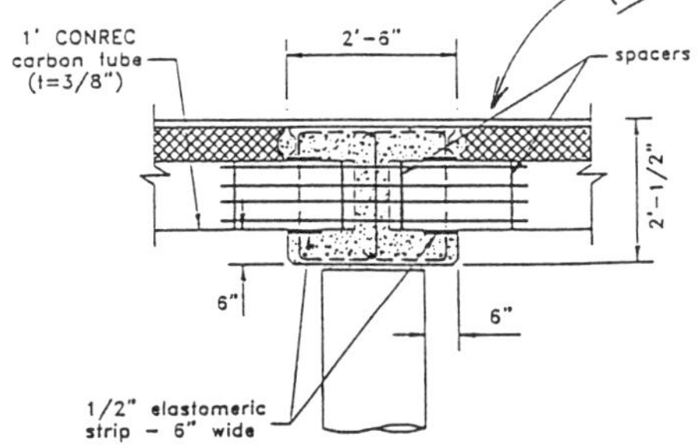

It seems obvious that, in the current United States economy, these composite materials are not competitive with the more common bridge materials now being used, unless accurate life cycle costs are considered. It would also seem obvious that the deck panels could solve the problem of deck deterioration in regions of heavy snow and ice where large amounts of salt are used for ice removal. These panels are nearly impregnable, they are not susceptible to corrosion, they are lightweight and do not require heavy lifting equipment like other materials, and they can be mass produced offsite to speed deck replacement. This could potentially be the largest market for advanced composites in bridge construction/rehabilitation in the US. In time, as usage increases the demand for higher quantities of these materials, the cost will lower and small bridges in remote locations can potentially be a good application.

The application to a large bridge with 100% advanced composites is still in the future on anything but an experimental basis, but that research and development must be continued to eliminate problems and reduce costs. A novel bridge across Interstate 5 in San Diego is being designed, under the ARPA-FHWA program, to connect two parts of the UC San Diego campus. The bridge is planned as an asymmetrical cable stayed four lane bridge to carry normal highway loading. Many of the components have been tested and are being utilized in the smaller bridges to gain better performance data under real traffic loading conditions.

We know the advantages of these advanced composite materials from the testing and field applications to date. We also know some of the obstacles to be overcome. These aprograms across the nation and especially the California program are designed to implement the use of these materials into bridge and highway infrastructure as research and good practice permit.

MATERIAL TESTING PROGRAM

The major concerns associated with the implementation of advanced composite materials into the civil infrastructure are long term durability and consistency in the field applications. It is imperative that we are able to consistently replicate in the field what we have tested in the laboratory. To insure the necessary quality control Caltrans, in conjunction with the Aerospace Corporation, has developed a comprehensive testing program for the evaluation of advanced composite materials for seismic retrofit and rehabilitation of structures.

The Caltrans program was set up to identify the critical parameters and procedures which need to be monitored or controlled to assure the reliable performance of composite retrofitted columns or bridge decks. Cost considerations are an important part of this program. It would be very easy to define tests, inspections, and quality checks that would increase the price of manufacturing composite jackets to the point where they would not be cost competitive with conventional materials. Because of the variations in composites, some testing is unavoidable. However, this program is designed to minimize the testing required to assure a quality product.

Design Issues And Durability Concerns

Caltrans' experience in composites research, trial field demonstrations, as well as through numerous meetings with the industry, revealed a myriad of issues that should be addressed by any public agency. Listed below are issues that must be verified by the engineer of record prior to using composites in infrastructure applications.

- Product documentation consistent with application
- Process Control
- Material Selection Criteria
- Material physical properties

- Long term durability (chemical and physical) testing of the composite against:
  - Moisture
  - Salt attack
  - Alkali attack
  - Ozone
  - High/low temperature extremes
  - Ultra violet
  - Other

- Quality control in manufacturing, mixing and applying
- Fiber content, voids, resin ratio
- Design guidelines for the specific composite
- Safety factors
- Damage and failure modes
- Adequate specifications
- Repeatability and consistency
- Acceptable field erection methods
- Effect of fatigue on bond behavior
- Performance under dynamic load
- Testing under sustained loading
- Qualifications of suppliers and product designers
- Cure temperature
- Transportation, handling
- Maintenance issues

Our experience has also revealed crucial issues that are unique to each fiber, resin, and equally important, the manufacturing process and application method. Composites material testing which was conducted by various research institutions show sensitivity to certain environmental factors and possible strength degradation. These results should not necessarily eliminate the use of composites in infrastructure; they merely underscore the need to properly select all components of the composite to suit applications and performance requirements. These results further show the need for safety factors of such magnitude typically not common in conventional construction materials.

In addition to column retrofit concepts, some manufacturers have tested upgrading structural members, such as beams and slabs, using carbon fiber. However, only empirical data was generated, with no significant design or durability guidelines. Even though the industry is rich in data related to aerospace and marine applications, the data we need, relevant to civil engineering infrastructure applications, is very limited.

Program Overview

The Caltrans program primarily focuses on two areas of applications:

1 - Seismic retrofit of bridges.
2 - Bridge strengthening and rehabilitation methods.

To ensure a sound objective technical evaluation, Caltrans is cooperating with several agencies which possess viable technologies, knowledge and tools to conduct a comprehensive assessment of the various systems under consideration. This cooperative effort is being facilitated by the Society for the Advancement of Material and Process Engineering (SAMPE). Material testing is being performed by the Aerospace Corporation (El Segundo, California). Structural testing is being conducted at the University of California at Irvine (UCI).

Program Objectives

Qualifying well documented composites materials and processes for structural applications is the ultimate goal of the Caltrans effort. In order to achieve such level of confidence, the Caltrans' program is set to accomplish the following objectives:

1 -Identify acceptable material testing methods appropriate for each material type (Carbon fiber, E-Glass, S-Glass, Aramid) and consistent with intended applications. This item includes identifying environmental and physical factors that must be addressed. This objective has been accomplished through the pre-qualification document.

2 -Identify and/or develop structural testing methods to verify shear, confinement and flexural strength of the composite system. The goal is to develop test methods that are capable of demonstrating the structural performance of a given system, yet simple and inexpensive. This objective has been accomplished.

3 -Develop analysis and modeling techniques appropriate for the intended application. Such analysis should take into account the interaction between the composite material and the structure. Work in this area is in progress.

4 -Establish performance criteria for the various materials.

5 -Develop standard specifications and necessary special provisions for viable systems. These specifications should address material types, manufacturing process, mixing and curing, quality control, quality assurance and application methods. Where applicable, ASTM tests will be identified and used. Several projects have been advertised already. Specifications were developed for those contracts.

6 -Develop and adopt design guidelines taking into account environmental and physical factors. Current design guidelines incorporate environmental factor of safety. This factor will be re-examined at the conclusion of the program for any possible need to adjust.

Material Testing

Caltrans issued its pre-qualification requirements in April 1996 and later amended such requirements in January 1997. During the same period, Caltrans issued its Memo-to Designers, which states the conditions under which composite alternatives may be used. To help industry participants qualify, Caltrans is carrying out this program for qualifying composite jackets for seismic retrofit of bridge columns. The Aerospace Corporation is supporting Caltrans in the qualification program and is performing environmental durability qualification tests. Degradation of mechanical and physical properties of composite panels is being determined following exposure to various environmental conditions for periods up to 10,000 hours. Environmental exposures include 100% humidity at 100oF immersion in salt water, immersion in alkali solution, ultraviolet light, dry heat at 140oF, a freeze/thaw test, and immersion in diesel fuel. The effects of the environmental exposures are being quantified by measurements of the composite panel mass, tensile modulus, strength, and failure strain, interlaminar shear strength, and glass transition temperature. Property measurements are being made after exposure intervals of 1,000 hours, 3,000 hours and 10,000 hours to allow estimates of degradation over the projected service life. As of December 1996, property testing following the 1,000 hours and 3,000 hours exposure periods has been completed for three glass fiber/polymer resin systems and for four carbon fiber/polymer resin systems.

Structural Testing

All composite column casing systems are required to satisfy reduced scale cyclic column testing requirements to verify the casing constructability and effectiveness as a seismic retrofit measure. To qualify a system as an alternative column casing for seismic retrofit, a minimum of two types of retrofit enhancements must be demonstrated and tested in accordance with Caltrans requirements. Test results must satisfy Caltrans requirements relative to ductility performance, shear strength, and flexural enhancement. For each shape, cyclic tests must be conducted to demonstrate the performance of both retrofit enhancements and corresponding unretrofitted "As-Builts". Manufacturers may elect to qualify only one shape (circular or rectangular) by satisfying all tests requirements for either the circular tests or rectangular tests, thus limiting their qualifications to these systems.

For each geometrical shape, and for each corresponding enhancement, a minimum of one retrofitted "As-Built" column and one unretrofitted column shall be built and tested. For example, to qualify a system for circular column retrofit applications, the following four test specimens must be constructed & tested:

1. Circular Shear As-Built Column (Unretrofitted)
2. Circular Lap Splice As-Built Column (Unretrofitted)
3. Circular Shear Retrofitted Column subjected to double bending load
4. Circular Lap Splice Retrofitted Column subjected to single bending load.

All column details must conform to Caltrans requirements. Retrofit jacket thickness (or fiber ratio) must comply with the current Caltrans design criteria, with proper scaling factors when applicable, and shall satisfy the following:

1. Minimum confinement stress of 300 psi in the lap splice and/or plastic hinging zone
2. Maximum material elongation of 0.001 in/in in the lap splice zone and 0.004 in/in in the plastic hinging zone
3. Minimum confinement stress of 150 psi and material elongation of 0.004 must be maintained elsewhere in the column with appropriate transition
4. Minimum displacement ductility for the retrofitted column of 8 to 12 is to be expected.

An expected concrete strength of 5000 psi at the time of testing and Grade 60 reinforcing steel shall be used, although Grade 40 is preferable when available.

Summary Of Program Tasks

The following briefly summarizes tasks which are used to develop the information necessary to qualify vendors to wrap bridge columns with composites for the purpose of seismic retrofitting. All of the data will be cataloged and the program will be managed under one of the tasks. The proposed work includes an analysis of a variety of designs, materials and application techniques to determine the internal stresses in the composite and the strength of the jacket. Two of the tasks involve extensive testing of the composite materials, to fill holes in the database and, using materials from previously wrapped test columns, determine the effect of weathering/aging. Techniques and specifications will be defined under the quality assurance task to guarantee that the vendor's products are consistent and of sufficient quality to fulfill their function. Under the nondestructive evaluation task, techniques will be developed to verify the quality of the jacket as well as the health of the concrete itself.

Task 1: Analytical Design Verification - Modeling

Objective: Conduct analytical modeling of selected sub-scale tests and estimate a critical flaw size. Help develop a simplified guide for designing composite jackets.

Deliverable: Internal stress analysis of selected sub-scale tests and critical flaw size estimation.

Task 2: Composite Properties Characterization

Objective: Develop specific requirements for manufacturing and testing composite jackets. Identify limits (e.g.. temperature and humidity) allowed during manufacture.

Deliverables: List of recommended test methods.
Recommended manufacturing methods and placards.

Task 3: Reduced Scale Test Column Verification

Objective: Determine the quality of the wraps on the test specimens and the resolution of the nondestructive testing techniques.

Deliverables: Nondestructive evaluation maps of selected sub-scale columns both before and after testing.
Comparison of test results to analytical models.

Task 4: Quality Assurance

Objective: Establish the basis for a plan to assure that composite retrofitted columns uniformly meet established performance requirements defined by Caltrans.

Deliverables: Define standard test procedures for incoming inspection and witness specimens. Specify/define minimum requirements for quality testing, e.g., number of witness specimens required.

Task 5: Non-Destructive Evaluation

Objective: Finalize and document column assessment techniques

Deliverables: Document the most effective NDE techniques.
Demonstrate techniques on sub-scale columns.

Task 6: System Evaluation

Objective: Develop a manufacturing model to compare total costs of composite jackets with steel jackets.

Deliverables: Estimation of labor and material costs for composite jackets and steel jackets
Life cycle cost estimates

Task 7: Database Organization and Project Management

Objective: Collect, assimilate, and store the generated data into a database. Manage tasks 1 through 6.

Deliverables: Management, schedule and cost reports Database generated by this and related programs including: material properties, NDE methods, manufacturing specifications, processes and modal studies.

Preliminary results are now available and are published in a report by Sultan of Caltrans and Steckel of Aerospace Corporation. More complete results will be available during the winter of 1999.

FIELD APPLICATION QUALITY CONTROL SPECIFICATIONS

Caltrans has developed preliminary construction specifications to ensure quality control for the field applications of advanced composite materials. Separate specifications are available for the various materials but they are generic enough to allow the various vendors of each material to bid, assuming they have passed the qualification tests. Design guidelines are also available for determining the proper thickness of materials.

SUMMARY

Caltrans has embarked on a program to utilize the advanced composite materials in seismic retrofit strengthening of bridge columns and other structural members. The goal is to increase the shear capacity and develop ductile performance in these members during a seismic event. It seems obvious that, in the current United States economy, these composite materials are not competitive with the more common bridge materials now being used, unless accurate life cycle costs are considered. We know the advantages of these advanced composite materials from the testing and field applications to date. We also know some of the obstacles to be overcome. These programs across the nation and especially the California program are designed to implement the use of these materials into the bridge and highway infrastructure as research and good engineering practice permit.

REFERENCES

1. California Department of Transportation, Draft Special Provisions-Section 10-1._ Alternative Column Casing, June 1997

2. Priestley, M.J.N., Seible, F., and Calvi, G.M., "Seismic Design and Retrofit of Bridges" John Wiley & Sons, 1996, 686 pp.

3. Seible, F., "Advanced Composites for Bridge Infrastructure Rehabilitation and Renewal" International Conference on Composite Construction, Innsbruck, September 1997

4. Steckel, G.L., and Sultan, M., "Evaluating Advanced Composites For Application In Transportation Structures-A Program Overview" Second FHWA/Caltrans National Seismic Conference, Sacramento, CA, July 1997

Design Concepts for Composite Plate Bonding and Column Confinement

L. HOLLAWAY

ABSTRACT

Over the past decade research has clearly demonstrated that reinforced concrete members can successfully be retrofitted with carbon fibre reinforced polymer to enable the structural units to overcome construction and degradation deficiencies. The advanced polymer composite material can readily be installed on site but the analysis and design of the retrofitted material in conjunction with the reinforced concrete must be undertaken using solid engineering design principles. This paper will discuss a method of design for plate bonding to the soffit of the beam and for the confinement of concrete for columns. It will illustrate the possible failure zones in a retrofitted RC beam; these areas must be considered in the design. A possible design for a confined concrete is also discussed.

1 INTRODUCTION

The need to strengthen concrete structures is a worldwide one which places considerable importance on rehabilitation methods particularly with regards to bridges for which much of the world's infrastructure was originally designed for smaller vehicles in less congested traffic. The increase vehicle loads and densities, combined with the recurrent use of de-icing salts, has insured that many structures are near to or beyond the end of their lives. In European Union alone, nearly 84,000 reinforced and pre-stressed concrete bridges require maintenance, repair and strengthening with an annual budget of £215M, excluding traffic delays and management costs Ref [1], All European highways should have been capable of carrying 40 tonnes. vehicles by the end of the millennium, this is an increase from the current requirement of 32 tonnes, in accordance with a European Community Directive, but some 40,000 bridges in the UK alone have failed to meet this requirement. Many structures built in the United States during the construction boom of the 1960's with little attention to durability issues and inadequate knowledge of seismic design are in need of urgent repair and retrofit. Earthquakes in urban areas of the United States (Loma Pricta 1989, Northbridge 1994) and in Japan (Kobe 1995) have demonstrated the inadequacy of old seismic design codes. Provision in seismic design codes, ageing and environmental deterioration are the

---

Professor Len. Hollaway, Department of Civil Engineering, University of Surrey, Surrey, UK.

primary reason that thousands of structures have been identified as substandard. The value of the Unites States infra-structure has been estimated at $20 trillion Ref. [2]; in 1996, 31% of the estimated 582,000 highway bridges in the US were identified as structurally deficient or functionally obsolete Ref. [3]. An average of $3.1 billion/year is spent on existing highway bridges, $1.3 billion of which is spent for minor and major bridge rehabilitation. The need, therefore, to strengthen concrete structures is economically paramount in the highway infrastructure, the rehabilitation of structural members in buildings is also of importance.

The behaviour of retrofitting systems depends upon the type of member that is to be retrofitted; typically these fall into three categories:
- flexural strengthening
- shear strengthening
- confinement strengthening.

The paper is concerned with the improvement of the live load carrying capacity of bridges and buildings and the increase of dead load in structures, from the point of view of the design philosophy and the design guide lines to be achieved for a successful rehabilitation of the structures.

2 THE DESIGN TECHNIQUES

In considering the design technique for flexural plate bonding, there are an number of design parameters which should be considered when optimizing the strengthening system. These are given in Ref. [4] and have been reproduced under:
- the plate width and thickness
- shear span to beam depth ratio for bond length considerations
- the bond length and anchorage considerations
- plate geometry
- adhesive thickness

The combination of these parameters is an important factor in design. For example, a specific reduction of mid-span deflection at, say, service load could be achieved by a variety of modifications, for instance, by changing the thickness and/or the length of the composite plate. However, the ultimate load could vary due to the fact that the failure mechanisms change from quasi-ductile to brittle.

These parameters have been discussed fully in Ref. [5], only the general observations and the characteristics of plate bonding will be examined here.
- The addition of the external FRP plate has a limited structural effect on the first cracking load. However, substantial increases do occur in post cracking stiffness, in the serviceability load, the load at which yielding of the internal steel occurs and the member stiffness after yielding. The maximum load may also be increased significantly, although the ductility to collapse, in comparison to an unplated member is generally reduced, especially if the plate ends are not anchored.
- The ultimate mode of failure will depend upon the configuration of the strengthening system, in particular whether anchorage at the plate ends is provided. If the plate ends are not anchored and the shear span to beam depth

ratio (a_v/h) is less than 3.0, failure will occur almost invariably by plate peel initiated by the occurrence of a shear crack at the free end of the plate; this will occur in a brittle manner.
- If the shear span to beam depth is in the range of 3.5 a shear bond separation of the plate under the action of a shear crack is likely to occur; this is shown in figure 1. For shear span to beam depth ratios greater than 4.0 a shear bond separation of the plate
- under the action of a flexural-shear crack is likely to occur; this is illustrated in figure 2. These observations are for like beams, for dissimilar beams other parameters may affect the failure mode.
- The adhesive thickness appears to have little apparent effect on the overall structural behaviour or on the stress distributions in the beam. However, it is generally considered that a 2 mm thickness is the optimum to enable any irregularities in level in the soffit of the beam to be accommodated and to fill any indents in the surface of the beam.
- The strengthening of a plated beam made from high strength concrete will result in greater relative increases of serviceability, yield and ultimate load, as well as post-cracking stiffness compared to the un-plated beam.
- As the failure of a plated beam is associated with a peeling of the plate from the beam, (although this is not exclusive as the plate can fail in tension) whether this is from the free end of the beam or by a shear crack step at an external load, the continuation of the plates up the vertical sides of the beam will tend to increase the resistance to the peel forces and will increase the ultimate load. There will, however, be little effect upon the ductility but the post cracking stiffness will be increased slightly due to the restraining effect of the side plates on cracks.

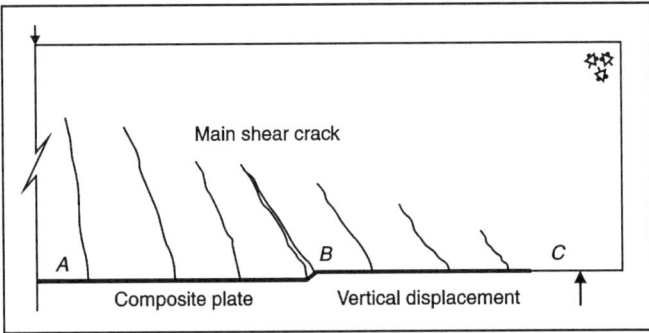

Figure 1   Initiation of plate separation for a beam shear span to beam depth ratio of 3.5

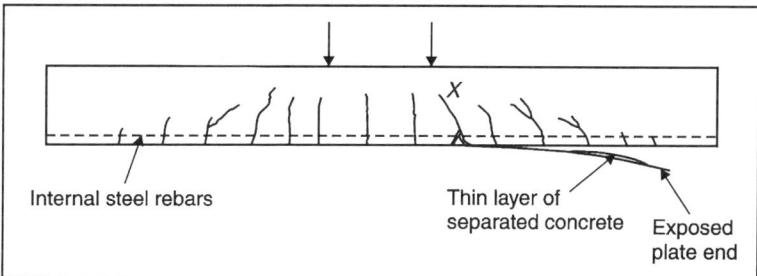

Figure 2 Typical mode of plate separation for a shear/span ratio of 4.0

3 THE FLEXURAL PLATE BONDING REINFORCEMENT REQUIREMENTS

The main requirements to be fulfilled by plate bonding are:
- a reduction of deflection of the beam (an increase in stiffness of beam)
- a limitation on the crack development in the concrete beam
- an increase in the load carrying capacity of the beam (serviceability load)
an increase in the load causing failure of the beam.

In order to achieve one or more of the above effects, the designer must consider carefully the design parameters listed above. Having completed this preliminary work and decided upon the thickness and width of the plate to be used, the thickness and type of adhesive and the bond length, the designer must consider failure modes of the beam system.

Central to the analysis and design of FRP strengthened RC members is the identification of possible failure modes. The classical failure modes of plated RC and PC systems is by one of the following:
Figure 3 shows the possible failure modes for a beam strengthened in bending by an FRP plate.

- steel yielding followed by FRP plate fracture, this implies that both the FRP and the steel reinforcement ratios are low where yielding of the steel is followed by tensile failure of the FRP, (area 2 figure 3),
- steel yielding followed by crushing of the concrete, this implies that the steel reinforcement ratios are low and the FRP ratio is relatively high, (area 2 figure 3),
- yielding of compression reinforcement, (area 3 figure 3),
- concrete crushing, this implies that the steel and the RP reinforcement ratios are relatively high and failure of the section occurs in a brittle manner, (area 1 figure 3),
- tensile failure of the composite, (area 4 figure 3),
- bond failure between the FRP reinforcement and the concrete, the tensile strength of the adhesive is invariably greater than that of the concrete.

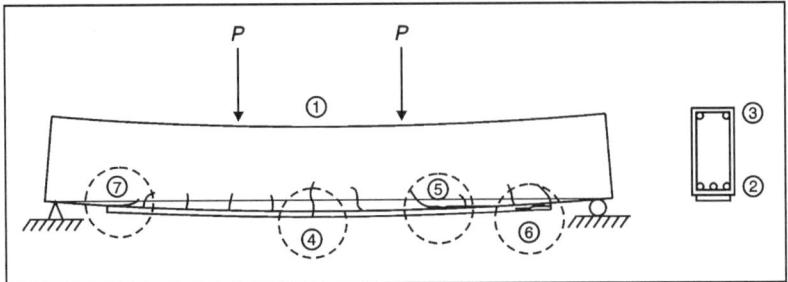

Figure 3 Possible failure modes for a beam strengthened in bending by an FRP plate

The first five failure modes are well understood but the bond failure between the FRP and the concrete does require further explanation.

The bond failure may occur at a number of different locations:
(a) in the anchorage zone due to the longitudinal shear and peel forces, (area 7),
(b) in areas of high moment due to stress concentrations at flexural cracks,
(c) in areas of high shear due to discontinuities at shear cracks, (area 5),
(d) in areas where the concrete surface is uneven and the laminate experiences high peel forces.

4 BASIC DESIGN GUIDANCE FOR ENGINEERS

As yet there are no National or International Standards dealing specifically with the design of strengthening concrete structures using fibre composite materials and adhesive joining technology. However, throughout the world there have been bodies set up to draw up guidelines and design rules for the technique; these committees sit in Europe, Britain, USA, Canada and Japan.

The European/UK procedures which are currently being developed will be used for the discussion in this paper to provide guidelines for non-pre-stressed FRP strengthening system.

References will be made to the Highway Agency documents BD44/95 – *The assessment of concrete highway bridges and structures,* BA30/94 *Strengthening of concrete highway structures using externally bonded plates.* BS 5400 *Steel, concrete and composite bridges* and BS 8110 *Structural use of concrete.*

Before deciding whether a structure requires strengthening the British Concrete Society has suggested that the following questions should be addressed.
- Has the condition or load carrying capacity decreased significantly?
- Has the loading changed significantly?
- What are the risks to the public, commerce and the structure?
- What are the cost implications of demolition and rebuilding?

TABLE 1   ULTIMATE AND SERVICEABILITY LIMIT STATE CHARACTERISTICS

Ultimate Limit State	Serviceability Limit State
Strength characteristics	Deflection characteristics Cracking characteristics Steel stress value Fatigue/creep characteristics Durability characteristics

- How would the strengthening works affect local infrastructure, commerce, safety, landowners and the environment?
- Are there any political issues involved?
- What is the age of the structure and is it of historical importance?
- What parties and authorities would be required to approve the works?
- Are there any programming or funding constraints? (eg. local authority annual budgets).

The design of FRP strengthening systems are generally based upon limit state principles. Its aim is to achieve an acceptable probability that the strengthened structure will perform satisfactorily during its life time.

- Limit state falls into two categories, - these are Ultimate and Serviceability Table 1.
- Ultimate limit state normally encompasses mechanisms that cause partial or complete collapse, it is the main component for design.
- Serviceability limit state corresponds to the state which principally affects the appearance or the proper performance of the structure.

There are a number of observations which are highly relevant when strengthening beam systems. These are:

- Provided good bond exists between the FRP and the concrete substrate, the structure will experience very fine distributed cracks with narrow crack widths.
- These crack widths must not exceed those recommended in BS 8100 BS 5100.
- Steel rebars should not yield in the service state (this requirement will prevent permanent deformations from developing).
- Fatigue and creep can be taken into account by lowering design stresses.
- Fire could be included in the above limit states but is generally considered an accidental load.

5   THE MECHANICAL PROPERTIES OF THE MATERIALS WHICH ARE ASSUMED IN THE DESIGN PROCEDURE

5.1 CONCRETE AND STEEL MATERIALS.

- the strength of the concrete to be used in the design equations should be the characteristic 28 day compressive cube strength or the worst creditable strength as defined in BD 44/95. Where the latter values are available only, modified values for the partial safety factors may be used.

Table 2 Stress strain curves of the individual materials associated with the strengthening system.

Trade name	Type of Material	Strength (tensile) MN/m^2	Modulus of elasticity GM/m^2	Thickness mm
CarboDur	Uni-directional carbon fibre plate	2400	150	-
Enforce	Carbon fibre plate	220-250	150	1.2-1.4
	Carbon fibre sheet	3900	240	-
Replark Type 30	Carbon fibre steet	3400	230	-

- the characteristic tensile strength of mild steel and high yield steel reinforcement are 250 and 460 MN/m^2 respectively. Both the steel types have a mean value of modulus of elasticity of 200 GN/m^2.

5.2 FRP MATERIALS

The strength of the FRP depends upon the type and percentage of fibre used. There are no agreed standard specifications for their manufacture and, therefore, all designs must be based upon the manufacturers values given in their literature. The properties required are generally the tensile strength, modulus of elasticity and elongation at failure.

Typical properties of some mechanical characteristics of a range of FRP strengthening materials taken from the manufacturers data are given in table 2

The equations which will be used later in the paper for the design of the strengthened systems are based upon :
- the rectangular parabolic stress/strain relationship for concrete in compression (BS8110) and the equivalent rectangular stress block for concrete (BS 8110) have been used
- the horizontal top branch relationship for reinforcing steel (BS 8110 Part 1) has been used
- a straight line response for the FRP has been assumed.

6.0 PARTIAL SAFETY FACTORS FOR LOADS

6.1 BUILDINGS

The appropriate safety factors for characterizing the dead and superimposed live loads would be taken from BS 8110 Part1 table 2.1 and these values are 1.4 and 1.6 respectively for the ultimate limit state. For the serviceability limit state the partial safety factors are both 1.0.

In addition to the partial safety factors applied to take account of the dead and live loads the designer will need to address the possibility of accidents such as fire collision and vandalism.

6.2 BRIDGES

The same range of safety factors will be required to be applied to the stresses for the FRP strengthening systems but the actual values of the partial safety factors may be different for the actual loading, the loading combination and the limit state under consideration. In BS5400 Part 2 table 1 the loads acting on bridges are divided into two groups, viz. permanent and transient. The former group include dead loads, superimposed dead loads and filling material loads. The transient loads include primary highway loads, the footway loads and the wind loads. These partial factors applicable to each load type for given load combination/limit state should be based upon values in table 1 of BS5400 Part2. or BD37/88 HA standard for highway loads in the UK.

7.0 PARTIAL SAFETY FACTORS FOR MATERIALS

The following points must be addressed when dealing with the materials involved in the strengthening techniques.
- the uncertainties associated with fibre composite materials and with joining technologies for plate bonding must be considered by applying the relevant factors of safety to the material properties.
- the change in material properties with time must be considered by applying relevant factors of safety to the material properties.
- the mechanical properties of the composites will generally be determined from the manufacturer's technical information sheets - the properties will be highly dependent up the method of manufacture - table 3 provides recommended values of partial safety factors for the different manufacture methods.

8.0 DESIGN STRENGTHS OF STEEL, CONCRETE AND FRP

Partial safety factors for the three materials will be required to assess their design strengths. Possible values have been given in table 4

Table 3 Recommended values of partial safety factors γ_{mm} (after reference 6).

Method of manufacture	Additional partial safety factor γ_{mm}	
	Fully post-cured at works	Not fully post-cured at works
Hand-held spray application	2.2	3.2
Machine-controlled spray application	1.4	2.0
Hand lay-up	1.4	2.0
Resin transfer moulding	1.2	1.7
Pre-impregnated lay-up	1.1	1.7
Machine-controlled filament winding	1.1	1.7
Pultrusion	1.1	1.7

Table 4 Partial safety factors for strength at the ultimate limit state

Materials	Partial safety factor (γ_m)
Steel reinforcement	1.05-1.15
Concrete in flexure or axial load	1.50-2.0
Concrete shear strength without shear reinforcement	1.25-1.15
Carbon fibre reinforced polymer	1.40
Aramid fibre reinforced polymer	1.50
Glass fibre reinforced polymer	3.50

9.0 DEFLECTION AND CRACKING

The deflections and crack widths in strengthened structures should be kept within the specified limits as laid down in BS8110 and BS5400 as appropriate.

10.0 FATIGUE

FRP materials can fail at stresses below their tensile strengths. To aviod this failure, the design stresses in the FRP should be limited by applying partial safety factors. Possible γ_m values for carbon and aramid FRP might be 1.75 and 2.15 respectively.

11.0 FRP STRESS RUPTURE (CORROSION)

If sustained loads are applied to the strengthening system, rupture of some composite materials may take place. It is, however, unlikely that carbon fibre polymer composites will fail due to this cause but glass fibre polymer composite must be kept to a level of 20% of ultimate to prevent stress corrosion taking place.

12.0 ADHESIVE

Conservative safety factors should be applied to the characteristic values for the material properties. The guide on partial safety factors, which has been drawn up by Institution of Structural Engineers (London), reflect the various factors for determining the overall material safety factor, this is given by:

$$\gamma_m = \gamma_{m1} \, \gamma_{m2} \, \gamma_{m3} \, \gamma_{m4} \, \gamma_{m5}$$

where the values of the various factors are given in table 5. For connections subjected to long term loading, the overall γ_m should not be less than 4.

Table 5 Recommended values for partial safety factors to be applied to adhesive properties, (taken from *A guide to the use of structural adhesives* (The Institution of Structural Engineers (London) 1999).

Source of the adhesive properties	γ_{m1}
Typical or textbook values (for appropriate adherends)	1.5
Values obtained by testing	1.25*
Method of adhesive application	γ_{m2}
Manual application, no adhesive thickness control	1.5
Manual application, adhesive thickness controlled	1.25
Established application procedure with repeatable and controlled process parameters.	1.0
Type of loading	γ_{m3}
Long-term loading	1.5
Short-term loading	1.0
Environmental conditions	γ_{m4}
Service conditions outside test conditions	2.0
Adhesive properties determined for the service conditions	1.0
Fatigue loading	γ_{m5}
Loading basically static	1.0
Adhesive subjected to significant fatigue loading	See table 6

* Where manufacturers supply guaranteed minimum properties, factor of 1.25 applied to the 'values obtained from testing' in the table may be reduced slightly, to 1.2. However, it should be noted that quoted values will generally apply to standard test conditions. If the in-service temperature differs significantly from that at which the tests were carried out, an additional factor is applied.

13.0 DESIGN OF MEMBERS IN FLEXURE

The following conditions should be met when strengthening beams in flexure:
(1) the maximum moment should be considered at critical points,
(2) the risk of peeling should be investigated at all FRP cut off points,
(3) the risk of de-bonding of the FRP and the concrete substrate should examined,
(4) the shear capacity of the section should be determined,
(5) the ductility of the strengthened member should be verified,

Table 6 Partial coefficient, γ_{m5}, for fatigue strength, taken from *A guide to the use of structural adhesives* (Institution of Structural Engineers, 1999)

Degree of inspection applications	Fail-safe applications	Non fail-safe
Periodic inspection, good access	1.5	2.0
Periodic inspection, poor access	2.0	2.5
No inspection/maintenance	2.5	3.0

(6) the serviceability limit states, eg. cracking, deflection, creep rupture should be checked for compliance.

The following analysis of the design of a strengthened system is based upon unpublished work of the Concrete Society, London. The design should provide for a compressive failure of the concrete or a tensile failure of the FRP. In addition, both types of failure should be preceded by yielding of the steel rebars. It should be possible to predict the actual mode of failure by comparing the design ultimate moment, M. with the balanced moment of resistance of the section, $M_{r,b}$. Thus when $M<M_{r,b}$ the FRP will reach its design strain before crushing of the concrete (case 1) and when $M>M_{r,b}$ the concrete will crush before the design strain of the FRP is reached (Case 2).

The amount of FRP (area A_f) required for the condition $M<M_{r,b}$ can be determined from

$$A_f = M_f / \sigma_{fd}\, z \qquad (1)$$

where
σ_{fd} = ultimate design stress in FRP = $\sigma_f / \gamma_m\, \gamma_{me}$
M_f = moment of resistance provided by the FRP.
$M_f = M - M_o$
M_o = moment of resistance of un-plated beam.
z = internal lever arm

The amount of FRP required for condition $M>M_{rb}$ can be determined from

$$A_f = F_f / \sigma_f\, \gamma_{fm} = F_f / 0.8\sigma_f \qquad (2)$$

where γ_{fm} = partial safety for material and F_f = tensile force in FRP plate.

In calculating the design stress in the FRP allowance should be made for the initial strain in the concrete at the time of strengthening.

The actual stress in the FRP is $= E_{fd}\, (\varepsilon_t - \varepsilon_o) = E_{fd}\, (\varepsilon_f)$

where ε_t = strain value at level of FRP sheet based upon a linear strain variation in the strengthened member under specific load.
and ε_f = strain in FRP and ε_o = initial strain.

13.2 BALANCED MOMENT OF RESISTANCE

The following assumptions are made when determining the balanced moment of the beam. BS 8110 gives the various reduction coefficients
- sections that are plane before bending remain plane after bending
- equivalent rectangular stress block for concrete at the ultimate limit state are used
- slip does not take place between FRP and the concrete
- tensile strength of concrete is ignored.

The moment of resistance $M_{r,b}$ for balanced failure
where $A_s' = 0$ (i.e. no compression reinforcement) is given by

$$M_{r,b} = (0.67\, f_{cu}/\gamma_{mc})b\, (0.9x)\, [z + (h-d)] - (f_y / \gamma_{ms})\, A_s\, (h-d). \qquad (3)$$

where $A_s' > 0$ (i.e. compression reinforcement) is given by

$$M_{r,b} = (0.67 f_{cu}/\gamma_{mc})b(0.9x)[z+(h-d)] - (f_y/\gamma_{ms})A_s(h-d) +$$
$$(f_y'/\gamma_{ms})A_s'(h-d') \quad (4)$$
where $z = d - (0.9x)/2$ and $x/h = 0.246$

The $(0.9x)$ is the depth of the concrete stress block and (x) is the depth to the N.A.

14.0 DESIGN RESISTANCE MOMENT OF UN-STRENGTHENED BEAMS

From BS 8110 Part 1 the design moment of resistance M_o of un-strengthened beams, assuming the equivalent rectangular stress block, is

When $A_s' = 0$

$K = M_o / f_{cu} bd^2$ ($= 0.15$ in the limiting case when $x=d/2$ and $z=3d/4$. then M_o will be supporting the greatest possible moment at that section)

$z = d[0.5 + \sqrt{(0.25 - K/0.9)}]$

$A_s = M_o / 0.95 f_y z$.

When $A_s' > 0$

$K' = 0.156$ If $K \leq K'$ compression reinforcement will not be required.

$K = M_o / f_{cu} bd^2$.

$z = d[0.5 + \sqrt{(0.25 - K'/0.9)}]$

$A_s' = (K-K')f_{cu} d b^2 / 0.95 f_y (d-d')$ if $(K-K')$ is zero or negative, no compression reinforcement is required.

$A_s = K' f_{cu} b d^2 / 0.95 f_y z) + A_s'$

15.0 DESIGN RESISTANCE MOMENT OF FRP STRENGTHENED BEAM.

The design moment of resistance M_r of the FRP strengthened beam:

when $A_s' = 0$ and $x =$ distance to neutral axis from top of beam

$M_r = (0.67/1.5) \, 0.9 \, f_{cu} \, b \, x \, [z + (h-d)] + (f_y/\gamma_{ms}) A_s (h-d)$

$= 0.89 f_{cu} b (d-z) [z+(h-d)] + (f_y/\gamma_{ms}) A_s (h-d)$ (5)

also $M_r = F_s z + F_f [z+(h-d)]$

where $F_s = (f_y/\gamma_{ms}) A =$ force in tensile steel

and $F_f =$ tensile force in the FRP material

when $A_s > 0$

$M_r = 0.89 f_{cu} b (d-z) [z+(h-d)] - (f_y/\gamma_{ms}) A_s (h-d) +$
$(f_y'/\gamma_{ms}) A_s' (h-d')$ (6)

Also $M_r = F_s z + F_f [z+(h-d)] + F_s' (0.45x - d')$ (7)

where $F_s' = (f_y'/\gamma_{ms}) A_s' =$ force in compressive steel.

16.0 FRP SEPARATION FAILURE

As mentioned earlier the strengthened FRP beam system can fail by six different failure modes. Dimensioning of the FRP beam should be made such that failure is governed by steel yielding followed by concrete crushing, with the FRP still in contact.

16.1 FRP PEELING AT THE ANCHORAGE ZONE

The peeling-off failure at the FRP curtailment is the most critical when dimensioning the FRP reinforcement. The approach which might be used to estimate the peel-off shear and normal stresses at the FRP curtailment. Expressions for these stresses have been derived by Ref [7] originally for steel.

$$\tau = [V_c + M \sqrt{[\ G_a / E_f t_f t_a\ ]}\,] [(\ t_f / I_{cr}) (h_f - x_{cr})] \quad (8)$$

and
$$\sigma = \tau\,[(3\ E_a\ t_f\,) / (E_f\ t_a\,)]^{1/4}$$

where
V_c = shear force at curtailment of the FRP
M = moment at a distance $h_f / 2$ from the support (for simply supported elements).
E_a = elastic modulus of adhesive, t_a = thickness of adhesive
G_a = shear modulus of adhesive
t_f = thickness of FRP
b_f = width of FRP, h_f = distance from FRP centroid to top of concrete
I_{cr} = moment of inertia of cracked section
x_{cr} = depth of neutral axis of cracked cross-section.

The peeling-off failure may be assumed to occur when the maximum principal stress equals the mean tensile strength of the concrete, f_{ctk} / γ_c

$$\sigma / 2 + \sqrt{[\sigma^2 /4 + \tau_2]} = f_{ctk} / \gamma_c$$

The peeling-off failure may be avoided by:
- providing a bond length at the end of the FRP plate equal to or greater than $f_{ftk}\ t_f /(f_{ctk} / \gamma_c)$ where f_{ftk} is the mean tensile strength in the FRP plate.
- transferring the entire tensile force ($A_f\ f_{ftk}$), in the FRP plate, to the concrete via. clamping.
- ensuring that $b_f / t_f \geq 40$ - a rule of thumb devised for steel plate bonding.

16.2 FRP DEBONDING

Debonding of the FRP due to sudden crack propagation at the concrete/FRP interface may occur when :
- vertical flexural cracks lead to the development of horizontal ones
- a step occurs when the two adjacent ends of a shear crack causes the plate to be uneven at that point and a high peel stress is developed

17.0 SHEAR STRENGTHENING

Externally bonded FRP laminates or fabric material is an effective technique to increase the shear capacity of reinforced concrete members. The FRP may be extended into the compression zone of the beam or may form a complete ring around the beam. Generally the FRP will be positioned so that the principal fibre orientation is either 45^0 or 90^0 to the longitudinal axis of the beam being strengthened.

The effectiveness of the external FRP shear reinforcement depends upon the mode of failure, this could be by either peeling-off through the concrete cover near the concrete/FRP interface or by tensile fracture of the FRP at a stress which may be lower than the FRP strength in uni-axial tension. This latter condition could be caused by stress concentrations at rounding corners or at debonded areas. Therefore, depending upon whether debonding or fracture of the system will occur first will depend upon :
- the bond conditions
- the available anchorage length
- the type of attachment at the FRP curtailment
- the FRP thickness

17.1 SHEAR STRENGTH

The design shear strength of reinforced <u>rectangular</u> concrete members, specifically beams, strengthened with externally bonded FRP reinforcement, V_{sd} is given by: $\quad V_{sd} = V_c + V_l + V_f \quad$ (9)
where V_c = shear resistance of concrete = $n_c \, b \, d$ (10)
$\quad V_l$ = shear resistance of the links = $A_{sv} (0.95 \, f_{yv})(d / s_v)$ [BS 8110] (11)
or $\quad = A_{sv} (f_{yv} / \gamma_{mc})(d / s_v) \quad$ [BD 44/95]
V_f = shear resistance of the FRP = $(0.9 / \gamma_f) \, \rho_f \, E_f \, e_{fe} \, b_w \, d \, [\sin \beta \, (1 + \cot \beta)]$ (Ref [8]) (12)
where n_c = the design concrete shear stress determined using Table 3.8 in BS 8119 Part 1.

A_{sv} = total cross area of links
s_v = spacing of links
f_{yv} = yield stress of links
A_f = area FRP shear reinforcement = $2 \, t_f \, w_f$
t_f = thickness of FRP material
w_f = width of FRP strips
ε_{fe} = maximum strain in FRP , (may be limited to 0.005 - Ref [8])
β = angle between FRP and longitudinal axis of the member
d_f = effective depth of shear reinforcement usually equal to d for rectangular sections and (d - thickness of slab) for Tee-sections.
d = effective depth of cross section of beam.
s_f = spacing between the centre line of FRP strips. (for continuous shear reinforcement $s_f = w_f$).
ρ_f = FRP shear reinforcement ratio
$= 2t_f / b_w$ for continuously bonded shear reinforcement of thickness t_f
or $\quad = (2t_f / b_w)(w_f / s_f)$ for reinforcement in the form of straps of width w_f at a spacing s_f.
b_w = minimum width of cross section of beam.

Figure 4 illustrates the vertical and inclined FRP strips for shear strengthening. The effective FRP strain can be approximated to:

$0 \leq \rho_f E_f \leq 1$ $\varepsilon_{fe} = 0.011 - 0.02(\rho_f E_f) + 0.01(\rho_f E_f)^2$ (E_f in Gpa)   (13)

$\rho_f E_f > 1$ $\varepsilon_{fe} = -0.0006(\rho_f E_f) + 0.0016$ (E_f in GPa)   (14)

The large value of ε_{fe} correspond to considerable opening of diagonal cracks to the extent that the shear strength of concrete shear-resisting mechanisms is reduced by aggregate interlock. (Ref. [8] suggested that the value of ε_{fe} may be limited to 0.005)

18.0 DESIGN OF CIRCULAR CONFINED COLUMNS SUBJECTED TO AXIAL LOAD

The requirements for strengthening concrete bridge supports using fibre reinforced polymer composites are:
(1) Bridge supports that fail assessment may be strengthened using FRP to increase their flexural and shear strengths. In the UK the strengthening schemes will comply with the requirements of BD 48/93, (the assessment of strengthening of highway bridge supports).
(2) The plastic methods of structural analysis should not be used for column strengthening using FRP.
(3) The FRP would normally be designed for a service life of 30 years. For circular columns circumferential (hoop) FRP wrappings can be used to;
- increase the concrete compressive strength
- increase the compressive strain in the outermost fibre of the concrete failure
- increase shear capacity
- enable the compressive strength of the axial FRP to be used in design.

The assumptions for the design of circular columns are:
- FRP material is linear elastic to failure
- FRP jacket is wound uniformly along the longitudinal length of the column

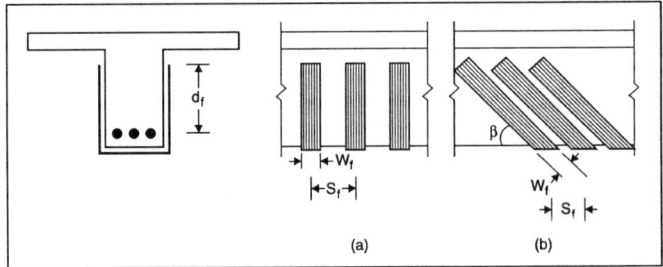

Figure 4   Vertical and inclined FRP strips for shear strengthening

- local cracking of expanding concrete during crack phase is negligible
- confinement pressure is constant in any section of the column.

Possible failure modes of a confined column are essentially:
- rupture of the FRP jacket overlap
- the tensile rupture of the FRP jacket.

For the first failure mode the circumferential (hoop) FRP shall satisfy the following:
- a maximum of two layers of FRP hoop fibres shall be provided
- an overlap shall be provided at all joins so that the FRP effectively acts as a continuous hoop. An overlap of $\geq$ 200 mm is recommended. eg. wrapping may consist of lengths of fabric of 2 x column circumference + 200 mm.
- hoop FRP shall be placed over axial FRP

For the second failure mode, which is the maximum performance of the system, the concrete in confinement will be completely cracked

The fracture of the column, under a compressive load, occurs by tension failure of the FRP. This is the ultimate confining stress obtained at the failure strength of the FRP.

Figure 5 shows a sketch of the wrapping of FRP around the column and the stresses across the column section.

$$f_r = (\rho_{cf} f_{cr})/2 \qquad (15)$$
$$\rho_{cf} = 4\, n\, t_f / D \qquad (16)$$

therefore $\quad f_r = f_t (n t_f)/R$

where
- f_r = ultimate confining stress
- ρ_{cf} = volumetric ratio of the FRP
- f_t = ultimate tensile strength of the FRP
- n = number of layers of FRP
- t_f = thickness of one layer of FRP
- D = diameter of core concrete confined with FRP
- R = radius of core concrete confined with FRP

The figure 6 represents a qualitative constitutive law in compression for circular concrete un-reinforced columns, reinforced columns with steel stirrups and concrete columns confined in an FRP jacket.

The main observations from this graph (figure 6) relating axial compressive stress to axial strain are as follows:
- the initial slopes of the graph are similar for the three columns
- the un-reinforced column fails at f_{co} after developing cracks

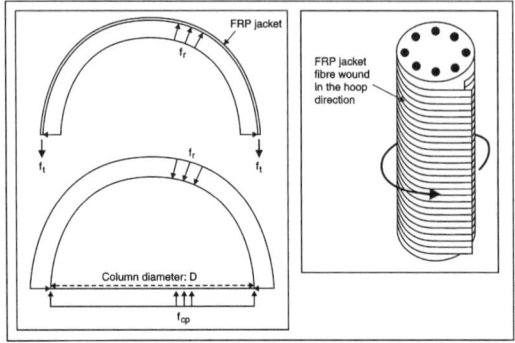

Figure 5 Sketch of wrapping of FRP around column (After Mbrace Composite Design Manual)

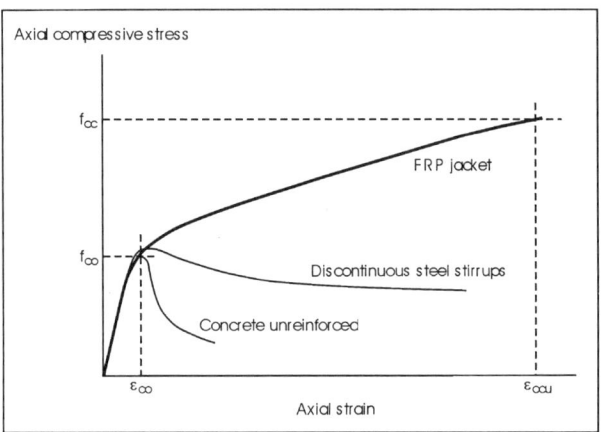

Figure 6. Typical constitutive law of concrete column in compression: un-reinforced, reinforced with steel stirrups and wrapped with FRP jacket

- the steel stirrup column presents a gradual decrease in strength at f_{co} with the formation of cracking

- the column on which the FRP confines the concrete, continues to take load. The stress-strain relationship is almost linear and the slope is directly proportional to the amount of GFRP material in the wrap.

Fracture of the contained column will occur by tensile failure of the FRP and the compressive strength of the contained concrete is given by the equation 17 (Ref.[9])

$$f_{cc} / f_{co} = 1.0 + 4.1 \, k_c \, f_r / f_{co} \tag{17}$$

Where f_{co} = compressive strength of plain concrete
F_{cc} = compressive strength of the wrapped concrete column

k_e = efficiency factor depending upon the type of material of the column, for concrete it is taken as 0.85.

It has been suggested in Ref, [10] that the entire constitutive law can be predicted by the bi-linear curve:

$$\sigma_c = [(E_1 - E_2) \varepsilon_c] / [1 + \{\varepsilon_c(E_1 - E_2) / f_{co} \}^n]^{1/n} + (E_2 \varepsilon_c) \quad (18)$$

where E_1 = the secant elastic modulus of the concrete
E_2 = the slope of the second branch of the constitutive behaviour in compression
f_{co} = compressive strength of plain concrete
n = an empirical factor that provides connection between the two branches of the constitutive law
= 8 (recommended)

$$E_2 = f_{cc} - f_{co} / \varepsilon_{cc} \quad \text{and} \quad \varepsilon_{cc} = \varepsilon_r / \nu_{co} [1 + \sqrt{\{(f_{co} r) / f_t (t_f n)\}}] \quad (19)$$

where ν_{co} = Poisson ratio of concrete in the elastic range
ε_r = maximum tensile strain in the FRP jacket

In the above approach to the ultimate capacity in pure axial load on the column, it is important to define a material partial safety factor to enable the long term loading of the confined concrete column to be taken into account. Generally the value of the partial safety factor will lie in the range 2 to 3.

For wrapped circular cross section columns, the contribution of FRP to shear capacity may be estimated. (Ref. [8])

$$V_{fd} = \pi D^2 [\rho_f E_f \varepsilon_{fe}] / [4 \gamma_f] \quad (20)$$

where E_f = FRP modulus of elasticity in the circumferential direction
D = diameter of column
ε_{fe} = maximum strain in FRP (may be limited to 0.005).
ρ_f = (2 n t_f / D)   (for continuous jacket)
ρ_f = (2 n t_f / D)(s_w / s)   (for discontinuous jacket)
t_f = thickness of one layer of FRP laminate, n = no. of laminates
w_f = width of straps s_f = spacing of straps.
γ_f = partial safety factor for FRP material

The Highways Agency in the UK have produced a draft Interim Advice Note on 'Strengthening concrete bridge supports using fibre reinforced polymers'. The design basically reduces to the designer employing a series of design charts.

18.1 DESIGN OF RECTANGULAR COLUMNS SUBJECTED TO AXIAL LOAD

The effectiveness of the confinement of the concrete column is dependent upon the cross-sectional shape of the column.

Equation 18 for the value of f_{cc}, for circular concrete columns, can be modified by including a coefficient k_f, which will range between 0 and 1. The value for f_{cc} will then be:

$$f_{cc} / f_{co} = 1 + 4.1 k_e k_f [f_r / f_{co}] \quad (21)$$

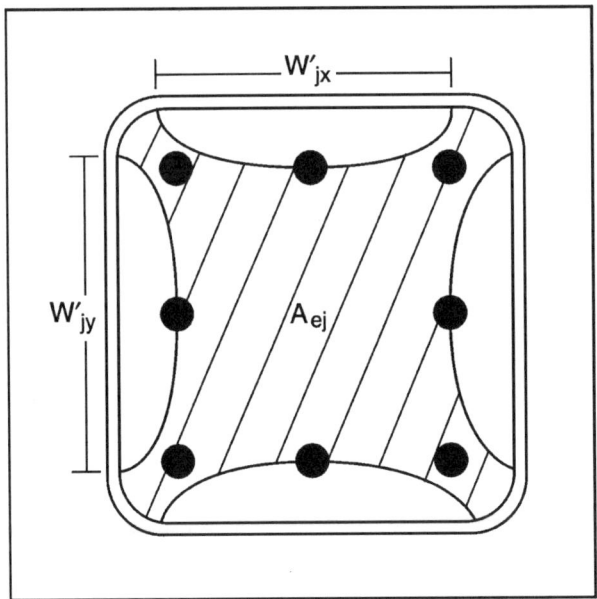

Figure 7 The effective area of confinement for a square column. (after MBrace Composite Design Manual)

By changing the shape of the cross-section of the column the confinement distribution will be modified. The confinement area in a square cross-section RC column is represented in figure (7). The confinement for this column will be in the zone near the corners, consequently, the effectiveness of confinement of square columns is lower than that for a circular column. Only the hatched area is effectively confined.

19 REFERENCES

[1] Lane, L S, Leeming, M B and Fashole-Luke, P S. (1997) 'Using *advanced composite materials in Bridge strengthening: introducing Project ROBUST,* The Structural Engineer, Vol 75, No. 1, pp16-18.

[2] NSF (1993) NSF 93-4 Engineering brochure on infrastructure U. S. National Science Foundation, Arlington, VA.

[3] FHWA (1998) *'Our Nation's Highways'* U.S. Department of Transportation, Federal Highway Administration, Publication No. FHWA-PL-98-015, McLean, Virginia.

[4] Hollaway, L C and Head, P R. 'Advanced polymer composites for construction', to be published by Elsevier Applied Science Publications.

[5] Hollaway, L C and Leeming, M (eds) (1999) *Strengthening of reinforced concrete structures using externally-bonded FRP composites in structural and civil engineering* Woodhead Publishing Ltd. Cambridge.

[6] Clarke, J L (ed) (1996) *Structural design of polymer composites -EUROCOMP design code and handbook.* London E&FN Spon p40.

[7] Roberts, T M (1989) *Approximate analysis of shear and normal stress concentration in the adhesive layer of plated RC beams, The Structural Engineer* vol. 67 No. 12 pp 228-233.

[8] Triantafillou, T C (1998) Shear strengthening of reinforced concrete beams using epoxy-bonded FRP composites *ACI Structural Journal* Vol. 95, No. 2 pp 107-115

[9] Miyauchi, K. Nishibayashi, S. and Inoue, S., E*stimation of strengthening effects with carbon fibre sheet for concrete column,* Proceedings of third International Symposium 'Non-metallic FRP reinforcement for concrete structures' , Japan Concrete Institute, Sapporo, 1997.

[10] ACI Committee 440 (1996), 'State of the art report on FRP for Concrete Structures' ACI440R-96, Manual of Concrete Practice, American Concrete Institute, Farmington Hills, MI.

British Standards Specifications issued by the British Standards Institution, London and Bridge Directives issued by The Highways Authority, London.

BD44/95 – *The assessment of concrete highway bridges and structures,* (Bridge Directive - Technical Note) issued by The Highways Agency, London, 1995

BD 48/93 – *The assessment and strengthening of highway bridge supports.* (Bridge Directive – Technical Note) issued by The Highway Agency, London, 1993.

BA30/94 *Strengthening of concrete highway structures using externally bonded plates.* (Bridge Advice Note) issued by The Highways Agency, London, 1994

BS 5400 Parts 1-5 : 1988 *Steel, concrete and composite bridges* British Standards Institution, London.

BS 8110. Parts 1-3 : 1997 *Structural use of concrete* British Standards Institution London.

Practical Implementation of Design Procedures for Retrofit of Bridge Columns Using FRP

R. A. IMBSEN and F. ALAMEDDINE

ABSTRACT

The application of Fiber Reinforced Polymer (FRP) casings has evolved as an alternative to steel column casings. Design procedures for FRP applications have been implemented to parallel procedures used in steel casings for seismic retrofitting of bridges. Explicit procedures using more complex methods have also been developed and employed in the design process. The objective of this paper is to outline the state of the practice in this field and shed some light on areas that differentiate FRP from steel casings applications.

INTRODUCTION

Bridge columns constructed in the United States prior to 1971 are generally deficient to resist seismic loads in shear, flexure and/or lateral confinement. Stirrups used were typically #4 bars spaced as 12 inches on center for the entire column length including the regions of potential plastic hinging. Typically, the footings were constructed with footing dowels, or starter bars, with the main longitudinal column reinforcement lapped onto the dowels. As the force levels in these columns approach yield, the lap splice begins to slip. At the onset of yielding, the lap splice degrades into a pin-type condition and within the first few cycles of inelastic bending, the load-carrying capacity degrades significantly.

Caltrans has developed three types of column casings as part of their comprehension bridge retrofit program. These types are shown in Figure 1 for steel casings. Those types are as follows [1]:
1. Type F Shell: this type provides a fixed end condition; footing retrofit may be needed.
2. Type P Shell: this type permits a pin to form by allowing lap splices to slip; footing retrofit is not required.

Roy A. Imbsen, P.E., Dr. Eng., Imbsen & Associates, Inc., 9912 Business Park Drive, Suite 130, Sacramento, CA 95827.
Fadel Alameddine, P.E., Ph.D., Office of Earthquake Engineering, Caltrans, 1801 30th Street, Sacramento, CA 95816.

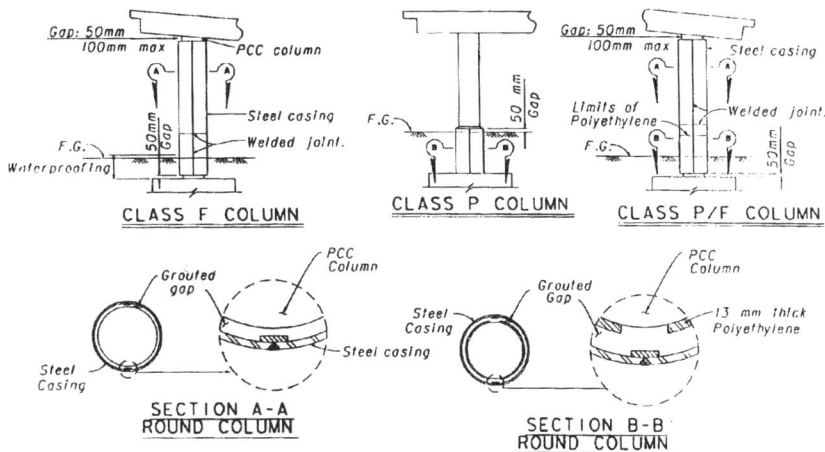

Figure 1. Types Of Steel Column Casing.

3. Type P/F Shell: this type is a full-length shell that provides a fixed condition at the top of column and has polystyrene, in the case of steel casing, at the lap splice to allow pin formation at the footing.

In multi-column bents, a pin condition is allowed at the base of the column, provided that fixity exists on top of the column and the shear capacity is adequate. The pin condition at the column base can be either an existing one or can be induced by bar slippage in a lap splice zone. By allowing a pin condition at the base of the column, footing retrofits are avoided on multi-column bridges typically.

For single-column bridges, a minimum of one Class F retrofit is recommended per frame. A frame is considered to be a part of the total length of a bridge containing more than one bent with an expansion joint at both ends. The rest of the columns in the frame may be modified with a Class P retrofit.

Using the different types of retrofit outlined above, the designer selects a retrofit strategy that best suits the dynamic characteristics of the bridge and leads to an economical and constructible design.

DESIGN OF STEEL COLUMN CASING

In California, steel jacking tests were conducted at University of California, San Diego (Priestley et al. 1996) and then implemented by Caltrans using some general guidelines that can be summarized as follows [2]:
1. Where a Class F retrofit is employed in the retrofit of a lap splice region at the base of a column the jacket needs to be designed for a confining pressure of 300 psi with a radial strain not exceeding 0.001. In the case of a circular column, the thickness of jacket "t" can be calculated as follows (see Figure 2):

$$t = \frac{P \times R}{f_s} \quad (1)$$

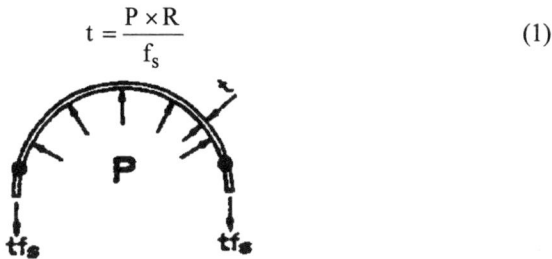

Figure 2. Jacket Hoop Stress Diagram.

where
f_s = stress in steel jacket, P = confining pressure, R = column radius

Because of the limiting strain of 0.001 in a lapped splice region:

$$f_s = \varepsilon E_s = .001 \times 29 \times 10^6 \text{ psi} = 29,000 \text{ psi} \quad (2)$$

In the case of continuous reinforcement in a plastic hinge zone, the stress in an A36 steel jacket is limited to a nominal value of 36 ksi.

f_s = 36 ksi

In summary, the casing thickness requirement for a Class F retrofit in a plastic hinge zone are calculated as follows:

$$t = \frac{R}{100} \times 12 \text{ in a region with lapped splice reinforcement} \quad (3)$$

$$t = \frac{R}{120} \times 12 \text{ in a region with continuous reinforcement} \quad (4)$$

2. For elliptical casings, the average column radius is used (see Figure 3).

$$R = \frac{B_x + B_y}{2} \quad (5)$$

The casing thickness "t" is calculated as shown above.
3. If t is calculated to be larger than 1", the designer should explore the use of anchor bolts, internal or external stiffeners to adequately confine the columns (see Figures 4 and 5).
4. Typically, Type P casing is 3/8" thick unless tall casings are used. In this case, a minimum ½" thickness is recommended for constructibility unless some measures are taken to prevent casing from bulging.

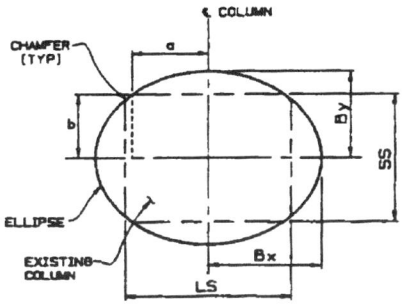

$$B_Y = \sqrt{b^2 + \frac{a^2}{(A_{SR})^2}}$$

$$B_X = B_Y \times A_{SR}$$

$$A_{SR} = \frac{LS}{SS}$$

A_{SR} = ASPECT RATIO
LS = LONG SIDE
SS = SHORT SIDE

COLUMN CASING
NO SCALE

Figure 3. Elliptical casing.

Figure 4. Use Of External Stiffeners on Steel Casings.

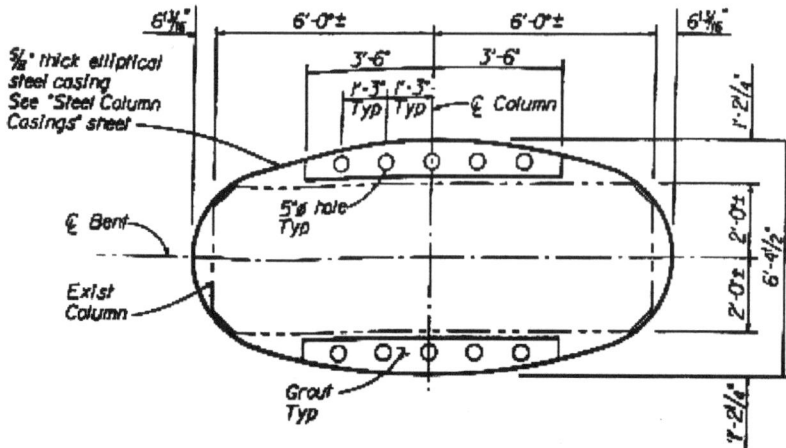

Figure 5. Internal Stiffener.

In general, once a Type F retrofit is identified in a column, the casing thickness is taken to be uniform over the full length of the column, unless tall columns are under consideration. For columns with nonuniform/unconventional sections, the typical guidelines are no longer applicable, especially if case (3) above applies. In this case, the designer has the responsibility of detailing a retrofit that adequately confine the column by means of using anchor bolts and stiffeners.

Experimental Findings of FRP Casings

The application of Fiber Reinforced Polymer (FRF) casings has evolved as an alternative to steel column casings. Recently there has been significant research and development using FRP casings to enhance column ductility [3&4]. Numerous tests were conducted on different types of composite casings, including Epoxy E-glass fiber, Epoxy resign-prepreg carbon fiber, prefabricated E-glass, and Epoxy carbon fiber. Discussion of lab results of these different types is beyond the scope of this paper. However, It is possible to discuss results of one set of tests conducted at UCSD to establish the wide range of applicability of composites on the retrofit of columns with different characteristics. This set of tests includes seven 40% scaled columns retrofitted using XXsys epoxy resin-prepreg carbon fiber [5-8]. Parameters of interest to structural designers are listed (see Table I) to establish some familiarity with limits of the tests that were conducted. These limits include:

a. The confinement pressure in plastic hinge zone at .001 dilation strain where lapped splice reinforcement exists and .004 dilation strain where continuous reinforcement exists.
b. The confinement pressure in shear enhancement zone at .004 dilation strain.
c. The shear stress applied to the column reported in terms of a multiplier of $\sqrt{f'_c}$.
d. The longitudinal column reinforcement ρ.

TABLE I. LAB TEST RESULTS USING XXSYS FRP CASINGS

Test	Column Longitudinal Reinf. ρ%	Applied Shear Stress $\times \sqrt{f'_c}$	Confinement Pressure			Mode of Failure
			Plastic Hinge Zone .001 Strain	Plastic Hinge Zone .004 Strain	Shear Enhancement Zone .004 Strain	
1. Carbon Fiber Jacket Retrofit Test of Circular Flexural Columns w/Lap Spliced Reinf.	2.5	2.5	300 psi	N/A	N/A	Rupture of starter bars at ductility 10.
2. Like Test 1 w/wrap thickness at 75%	2.5	2.5	225	N/A	N/A	Debonding in the lap splice region.
3. Carbon Fiber Jacket Retrofit Test of Circular Shear Bridge Column	2.5	5.0	N/A	360	90	Strength degradation and permanent opening of flexural cracks in the gap region at the bottom of column at ductility 10.
4. Seismic Retrofitting of Squat Circular Bridge Piers w/Carbon Fiber Jackets	1.5	6.5	N/A	900	120	Stable behavior up to ductility of 14 when jacket splits and interface cracks opened as much as 3/8".
5. Carbon fiber Jacket Retrofit Test of Rect. Column with Lap-Spliced Reinf.	2.5	2.3	338	N/A	N/A	Rupture of starter bars at ductility of 8.
6. Rectangular Carbon Fiber Jacket Retrofit Test of Shear Column	2.5	6.5	N/A	288	72	Carbon jacket bulged significantly at ductility of 8.
7. Rectangular Carbon Fiber Jacket of Flexural Column	5.0	4.8	N/A	600	60	Yield penetration in footing at ductility 6. Jacket rupture initiated by column bar buckling at ductility 8.

Tests 1 through 4 for circular columns show good results for a range of column shear stress and longitudinal column reinforcement. Class F retrofit for a lapped splice region is reasonably achievable. Test 5 shows the need of an elliptical casing for a Class F retrofit of a rectangular column with a lapped splice zone.

Tests 6 and 7 are for rectangular columns. The effectiveness of confining rectangular columns with rectangular jackets decreases significantly since only corner forces are generated during the dilation of the column flexural hinge. Tests on rectangular columns retrofitted with rectangular carbon jackets indicated a jacket efficiency of only 50% of that provided by a circular or an oval jacket. However, only column with side aspect ratio of 1.5 were tested. Thus, for columns with side aspect ratio of 1.5 or less, a jacket thickness twice the one calculated for an equivalent circular jacket should be used, whereas for columns with aspect ratio of greater than 1.5 extrapolation of test results is not recommended.

Tests 6 and 7 illustrate the fact that higher confinement pressure (i.e., larger thickness) does not indiscriminately improve the behavior.

The above observations are consistent with other types of FRP casings.

Design of FRP casings using Caltrans Method (Implicit Method)

Caltrans has currently seven FRP approved systems for column casings (see Table II).

Several projects have been completed in California using these systems. Some of these projects include:
- San Elijo Lagoon Bridges in San Diego, using epoxy resin-prepreg carbon fiber on multi-column bents (see Figure 6).
- Griffith Park Bridge in LA using Epoxy E-glass fiber on single-column bents (see Figure 7).
- Yolo Causeway Bridge in Sacramento using prefabricated E-glass to encase lap-spliced region of the pile extension (see Figure 8).

The design methodology is based on using tabulated values for jacket thickness depending on the system under consideration (see Figure 9). Based on Caltrans Memo to Designers 20-4, FRF casings are to be used on multi-column bents where fixity to the superstructure is achieved by continuous vertical reinforcement of the column extending into the superstructure [1]. Wrapping of the lapped splice region is not intended to achieve fixity of the column to the footing and is only intended to contain the degraded concrete at the base of the column. In this context, FRP casings are not used for retrofit of single-column structures.

The tabulated values for jacket thickness are derived based on achieving or confining pressure of 300 psi in the plastic hinge zone with a radial strain in the jacket corresponding to .004. For regions outside the plastic hinge zone, the criteria is reduced to a confining stress of 150 psi at a radial dilating strain of .004.

The expected performance level for columns with FRP casings is for ductility not to exceed six for circular columns and three for rectangular columns. Exceptions for ductility demands up to eight may be granted in special cases.

TABLE II. CALTRANS FRF APPROVED SYSTEMS

System	Manufacturer	Type of Composite Casing	Tensile Modulus 72 ±2°F × 103 (ksi) min.
1	Fyfe Co.	Epoxy E-glass fiber	3.7*
1	Hexcel Corporation	Epoxy E-glass fiber	3.7*
2	XXsys	Epoxy resin-prepreg carbon fiber	14.2*
3	Hardcore DuPont	Prefabricated vinylester E-glass	4.6
4	CMI	Prefabricated polyester E-glass	4.8
5	Master Builders-Mbrace	Epoxy carbon fiber	29.2*
6	Mitsubishi/Replark	Epoxy carbon fiber	31.2*
7	Mitsubishi/Toray	Epoxy carbon fiber	28.8*

*Tensile modulus of primary fibers

Figure 6. Epoxy Resin-Prepreg Carbon Fiber at San Elijo Lagoon Bridge.

Figure 7. Epoxy-E-Glass Fiber at Griffith Park Bridge.

Figure 8. Prefabricated E-Glass at Yolo Causeway.

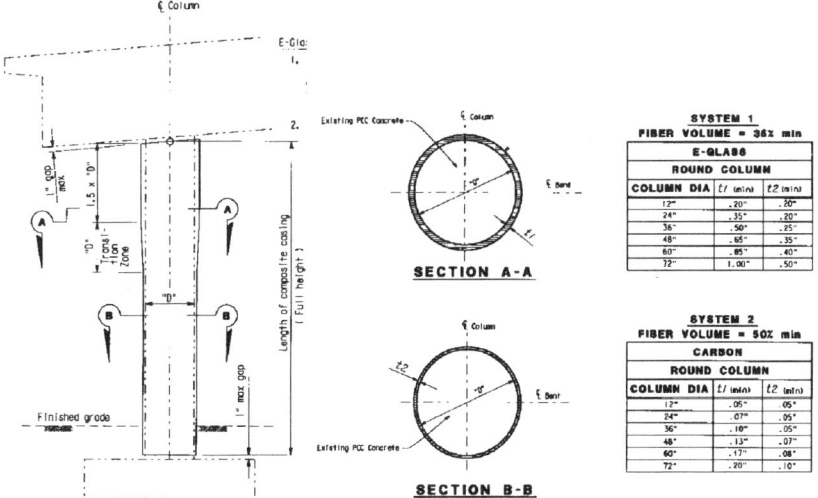

Figure 9. Design Aids using Caltrans MD 20-4.

Additional constraints are exerted on all FRP systems to be used. The following conditions are to be met for FRP to be specified:
1. For rectangular columns, the longest dimension is limited to a maximum of 36 inches. Rectangular column sides aspect ratio shall not be greater than 1.5.
2. For circular columns, the diameter must be 72 inches or less.
3. The total axial load (dead load + overturning) is not greater than 0.15 $f_c A_g$. Where f_c = concrete strength and A_g = column gross area.
4. The columns longitudinal reinforcement ratio is not greater than 2.5%.
5. The bridge does not require flame-sprayed plastic (i.e., not subject to projectiles in streams).
6. The columns must be prismatic in shape or have a constant, gentle taper.

Based on the description of completed projects and the guidelines listed above, it looks as though the above criteria has not always been followed to the letter since implementation of these new systems is actually an ongoing process.

Design of FRP Casing using an Explicit Method

The use of FRP casing is gaining momentum nationwide. The seismic assessment of a structure encompasses finding out the ductility demands on different members given a certain seismic event. This seismic event can be determined probabilistically or deterministically. Because of the variance in seismic input and the complexity of behavior of different structures with different levels of ductility demands, the use of an explicit method that is general in nature becomes more attractive.

This explicit approach involves the following:
1. Target a desired displacement ductility level on a given column
2. Use a confined concrete model such as provided by Mander et al or Pantelides et al, to determine the ultimate concrete strain ε_{cu} given a certain confinement level of the FRP casing [9&10].
3. Perform curvature analysis on given cross-section to calculate ultimate curvature ϕ_u and yield curvature ϕ_y.
4. Calculate plastic hinge length.
5. Calculate plastic displacement capacity Δ_p.
6. Calculate yield displacement Δ_y.
7. Calculate ultimate displacement capacity Δ_u

$$\Delta_u = \Delta_y + \Delta_p \qquad (6)$$

8. Calculate displacement ductility capacity

$$\mu = \frac{\Delta_u}{\Delta_y} \qquad (7)$$

9. If μ is equal to target value established in Step 1, then the iteration is over; otherwise go back to Step 2.
10. Check shear capacity of section taking into account the effect of FRP casing established in Step 2.

This approach has been taken in designing the retrofit/rehabilitation of the State Street Bridge at I-80 in Salt Lake City (see Figure 10). The retrofit of this model bridge is to be replicated on similar bridges along the I-80 corridor. Three design earthquakes were considered in the retrofit effort. Those include:
   a. 0.2g earthquake event
   b. An event having 10% probability of exceedance in 50 years.
   c. An event having 10% probability of exceedance in 250 years.

Several aspects of the retrofit of this bridge extended beyond the typical scope that is considered in the Caltrans implicit approach. These aspects include:
1. Casing of the lap splice region at the base of the column to ensure continuity to the footing.
2. Retrofitting of the column base to ensure lateral resistance of the bent in the longitudinal direction since the steel girders of the superstructure are only resting on top of the bent.
3. Extending the FRP retrofit to the joint region and the cap beam to ensure that the ductility levels anticipated during the earthquake event can be tolerated (see Figure 11)

In summary, the displacement-based explicit approach is warranted because it addresses the specific vulnerabilities of the subject bridge for different seismic events. The performance level is determined taking into account the capacities in all components of the subject bridge.

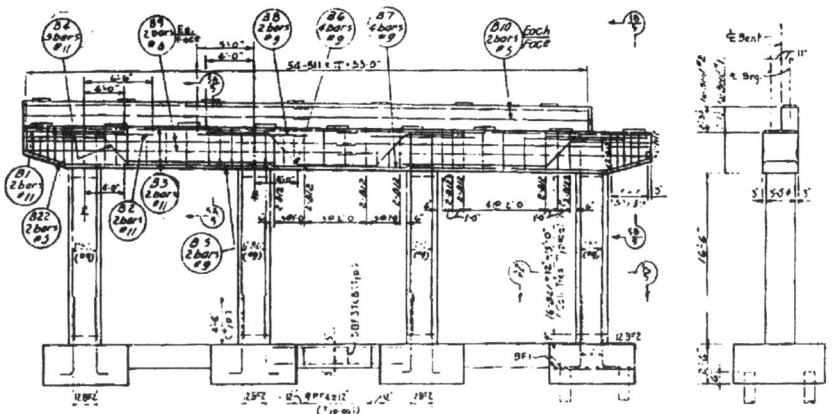

Figure 10. State Street As-Built.

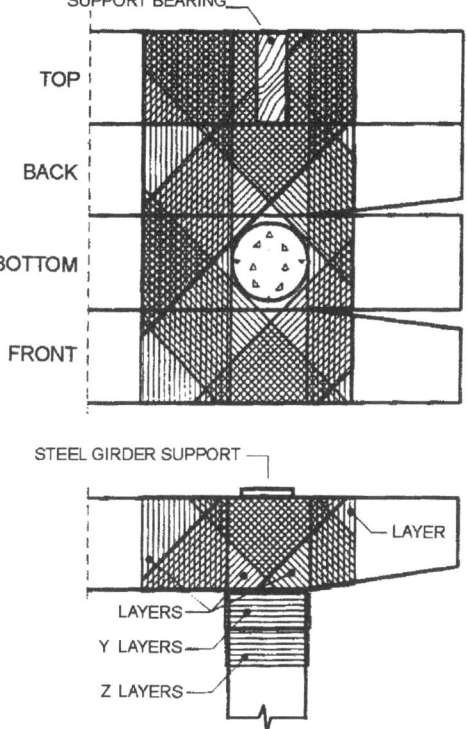

Figure 11. Part Elevation FRP Schematic Retrofit of State Street Bridge.

CONCLUSIONS

Both implicit and explicit design procedures are being used for the design process. Implicit design involves the use of design aids and tables.

ACKNOWLEDGMENTS

The authors would like to recognize the efforts of Mr. Sam Musser of the Utah DOT, Dr. Chris Pantelides of the university of Utah, Mr. Tom Sardo of the Iowa DOT, and Mr. Bill Gedris of the FHWA for their efforts in advancing the use of FRP.

REFERENCES

1. California Department of Transportation Memo to Designers MD 20-4.
2. Priestley, M. J. N., F. Seible, and G. M. Calvi. 1996. *Seismic Design and Retrofit of Bridges*. John Wiley & Sons, Inc., New York, NY.
3. Pantelides, C. Pl, I. Gergely, L. D. Reaveley, and R. J. Nuismer. 1997a. "Rehabilitation of Cap Beam-Column Joints with Carbon Fiber Jackets." *Proc., 3rd Int. Symp. On Non-Metallic (FRP) Reinforcement for Concrete Struct.*, Vol. 1, pp. 587-595.
4. Seible, F., M. J. N. Priestley, G. A. Hegemier, and D. Innamorato. 1997. "Seismic Retrofit of RC Columns with Continuous Carbon Fiber Jackets." *J. Compos. for Constr.,* ASCE, 1(2), pp. 52-62.
5. Carbon Fiber Jacket Retrofit Test of Circular Shear Bridge Column, CRC-2, Report No. ACTT-94/02, University of California, San Diego.
6. Seismic Retrofitting of Squat Circular Bridge Piers with Carbon Fiber Jackets, Report No. ACTT-94/04, University of California, San Diego.
7. Carbon Fiber Jacket Retrofit Test of Rectangular Flexural Column with Lap Spliced Reinforcement, Report No. ACTT-95/02, University of California, San Diego.
8. Rectangular Carbon Jacket Retrofit of Flexural Column with 5% Continuous Reinforcement, Report No. ACTT-95/03, University of California, San Diego.
9. Mander, J. B., M. J. N. Priestley, and R. Park. 1988. "Theoretical Stress-Strain Model for Confined Concrete." *J. Struct. Engrg.*, ASCE, 114(8), pp. 1804-1849.
10. Pantelides, C. P., M. W. Halling, K. C. Womack, L.D. Reaveley, I. Gergely, and R. M. Moyle. 1997b. "Carbon Fiber Composites for Rehabilitation of bridge Bents." *Proc., 2nd Symp. On Practical Solutions for Bridge Strengthening and Rehabilitation, BSAR II*, pp. 283-292.
11. Carbon Fiber Jacket Retrofit Test of Circular Flexural Columns with Lap Spliced Reinforcement, Report No. ACTT-95/04, University of California, San Diego.

Seismic Response of Reinforced Concrete Moment Connections Repaired and Upgraded with FRP Composites

A. MOSALLAM, P. CHAKRABARTI, S. SIM and H. M. ELSANADEDY

ABSTRACT

This paper presents the results of the second phase of a comprehensive research program at California State University at Fullerton on the applications of Fiber Reinforced Polymer (FRP) composites and adhesives in repair and retrofit reinforced concrete moment frame joint subassemblages. Primary focus is on the low reversal cyclic fatigue behavior of interior beam-column reinforced concrete joints. A description of this phase of the ongoing research program is presented. In this program, a total of twelve full-scale reinforced concrete connection tests were performed. Results obtained from the latest six specimens repaired and rehabilitated using REPLARK composite system are presented. Two connection specimens were used as control specimens and were tested to failure under inelastic reversal cyclic loading regime. The two "repairable" damaged specimens were re-tested under a similar loading regime after being repaired with both epoxy injection and carbon/epoxy laminates. To investigate the performance of the composite systems as retrofit systems, two other full-scale tests were conducted on undamaged specimens strengthened with specially designed carbon/epoxy laminates. P/δ and M/Θ hysteresis curve for each connection and the associated mode of failure are presented. Discussion on the advantages of using carbon epoxy laminates and recommendations for future research are presented. The results from a simple analytical model describing the behavior of connections retrofitted with composites is presented. The results of the analytical model were in a good agreement with the experimental results.

INTRODUCTION

The major influence of beam-column connections on the structural integrity and seismic performance of reinforced concrete structures has more evident after the 1989 *Loma Prieta*, the 1994 *Northridge* earthquakes, the Kobe earthquake of Japan, and the August 1999 Earthquake in Turkey. Post earthquake reports of the Loma Prieta, indicated that one of the main reasons behind the collapse of the Cypress Viaduct bridge, and the damage of the *China Basin* (see

---

A. Mosallam, P. Chakrabarti, and S. Sim, California State University, Fullerton, Fullerton, California, 92834, USA
H. Elsanadedy University of California, Irvine, Irvine, California, 92697-2175, USA

Figure (1-a)), and the *I-80 Nemit Freeway* is the failure of connections. A site survey, conducted by the author, of several parking structures in the Los Angeles in January 1994 following the Northridge earthquake, indicated that collapse of several portal frame structures were mainly due to the failure of beam-column and column-base connections (see Figure (1-b)).

 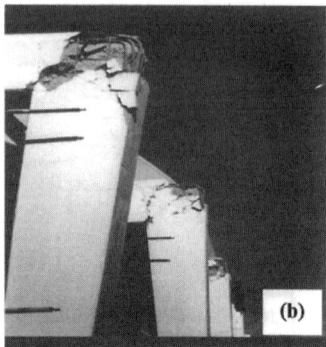

Figure (1): Joint Failure: a) I-280 China Basin Viaduct [Loma Prieta, 1989] b) Knee Frame Connection Failure [Northridge Earthquake, January 1994].

In the past few years, polymer composites were considered for several seismic rehabilitation and repair applications. This includes reinforced concrete (columns, beams, slabs, and walls), wood structures, steel members, and both reinforced and unreinforced masonry walls. A comprehensive review of different repair and rehabilitation applications using composites is presented in Ref [1]. The use of composites in seismic repair and rehabilitation of reinforced concrete joints was reported recently by Mosallam et al. [2], and Issa et al. [3].

NOMINAL CONNECTION STRESSES

It is common to model the actual stress distribution by a state of uniform stress. In general, the connection stresses can be computed by defining the connection volume by a width equal to the full width of the section, a height equal to the distance between the centroids of the beam bottom and top longitudinal reinforcement, and a length equal to the distance between the extreme layers of column longitudinal reinforcement as shown in Figure (2). As shown in the figure, axial tension and compression, shear and moments are acting on the connection from the adjacent or framing members (e.g. beam and column members). It is assumed that both tension and compression forces are acting at the centriod of the extreme longitudinal reinforcement, with values determined from the equilibrium of applied moments and the axial forces developed at the connection faces.

There are four main requirements for interior connection according to ACI-ASCE Committee 352. These requirements are:

1. Connection shear stress must be less than $20\sqrt{f_c'}\,(psi)$,
2. The minimum anchorage length for beam and column rebars is $20\,d_b$ or:

$$l_{dh} = 2.5 f_y d_b / 65\sqrt{f_c'} \, (kPa) \quad \text{[For beam rebars]} \tag{1}$$

where d_b= bar diameter (in); l_{dh}; f_y = steel reinforcement yield stress, and f_c' = concrete compressive strength in psi,

3. If rectangular hoops are used, the connection must contain a minimum transverse reinforcement equal to:

$$A_{sh} = (0.3 S_h h'' \frac{f_c'}{f_{yh}})(\frac{A_g}{A_c} - 1) \tag{2}$$

but not less than:

$$A_{sh} = 0.09 S_h h'' \frac{f_c'}{f_{yh}} \, \text{(ASCE-ACI 352)} \tag{3}$$

or

$$A_{sh} = 0.12 S_h h'' \frac{f_c'}{f_{yh}} \, \text{(ACI 318)} \tag{4}$$

where:
A_{sh} = Total cross-sectional area of transverse steel (in^2),
S_h = Reinforcement Spacing (in),
h'' = Confined cross-sectional dimension measured center to center of confining reinforcement (in), and
f_{yh} = yield stress of confining steel (psi).

4. The ratio of the sum of the flexural capacities of the columns (M_C) to the nominal moment strength of the beams (M_b) must be greater than 1.4 (ASCE-ACI 352) or 1.2 (ACI 318). In these computations, an overstrength factor a= 1.25 for the beam steel reinforcement should be included.

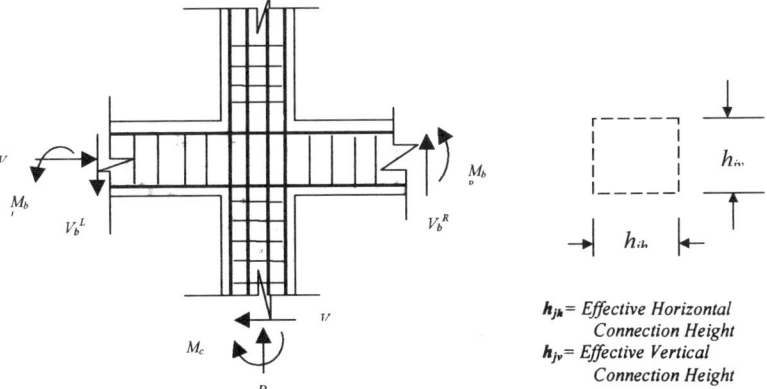

h_{jh}= Effective Horizontal Connection Height
h_{jv}= Effective Vertical Connection Height

Figure (2) Idealization of Connection Loads and Area Used in Nominal Connection Shear Stress Calculations.

The above requirements set forth by the ASCE-ACI 352 are to be used for new construction. However, the engineer will have difficulties to meet these requirements since this will require large column sections, deep beams, as well as large percentage of transverse steel reinforcement in the connection congested area. It should be also noted that there are no current design guidelines for repair and rehabilitation of existing beam-column joints given by ASCE-ACI 352. The current project aimed at providing innovative methodology for repair and rehab of existing connections

STRUCTURAL PERFORMANCE OF BEAM-COLUMN JOINTS REPAIRED & RETROFITTED WITH COMPOSITES

General: A total of six large-scale connections were tested under full reversed cyclic loading conditions. Dimensions and reinforcement details are shown in Figure (3). As shown in this figure, no special reinforcement details were specified at the joint region. This arrangement simulates old existing building constructed in the late 1950's and 1960's that would be good candidates for external repair and rehabilitation using REPLARK® composite system. The specimens were tested in an existing multi-propose testing frame spanning 30 feet (9.15 m) and which is equipped with two hydraulic actuators with total cyclic capacity of ±100,000 lbs. (445 kN) at the SRRS Center of CSUF. The testing frame structure is connected directly to a state-of-the-art data acquisition system, which was used to collect load, deflection, and strain information. In order to obtain complete moment-rotation histories, four LVDT's were used to continuously measure the relative rotation between the column and the beams for each connection specimen. In addition, a number of strain gages were placed at several critical locations at the steel reinforcements, concrete surfaces, and at the extreme fibers of the composite laminates in different directions.

Loading History: Figure (4) shows the test setup used in this program. The effect of an earthquake was simulated by applying reversed cyclic loading to the tip of the vertical member of the joint. The load and displacement histories were divided into two phases. Initially, the tests were conducted under a *load control* mode until the yielding load of the connection' steel reinforcements. At this stage, a *displacement control* mode was utilized. The loading regimes used in this study were adhered to the ICBO AC125 requirements.

Specimens' Lay-up: As it was mentioned earlier, two layup were used in this program. The first layup was simulating the horizontal member as a beam while the vertical member as a column. In this case, it is impractical to wrap around the beam due to the interference of the floor slab and the only possible way is to apply the REPLARK® sheets in a U-shape around the bottom and the two sides of the beam (Figure 5-a), while wrapping is possible for the column (vertical member) as shown in Figure (5). The other layup assumes that the horizontal member is the column and the vertical member is the beam (floor seismic loads are in this case simulated by the vertical applied cyclic loads). In this case, wrapping the column (horizontal member) is possible, while wrapping the vertical member (in this case a beam) is impractical as shown in Figure 5-b.

Control Specimen C1: The first connection control specimens was tested under both load-control (up to steel yielding) and displacement control regimes (up to failure). The P/δ Hystriseses for Control Specimen

C1 is shown in Figures (6). As shown in this figure, the ultimate load of specimen C1 was 21.83 kips (97 kN) which is equivalent to 469.35 kip-inch (53 kN-m). The ultimate vertical displacement and rotational angle for the push and the pull cycles were 1.46" (37 mm), 0.028 radiant, and 1.72" (43.7 mm), 0.035 radiant, respectively.

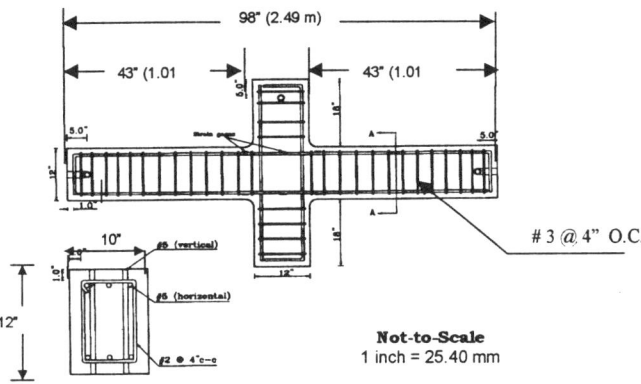

Figure (3): Dimensions and Reinforcement Details of Moment Frame Connection Specimens.

Figure (4): Interior Connection Test Setup

Repaired Specimen RU: The control specimen C1, described earlier, was removed from the testing frame, and was transported to the laboratory backyard for repair. All loose concrete was removed, and surfaces were cleaned using compressed air. After applying a bonding agent to the concrete surfaces, a high strength epoxy mortar was then used to fill all the gaps, and was left to cure for 24 hours. All cracks were then injected with low viscosity epoxy. After allowing the recommended cure time, the concrete surface was sand blasted, and cleaned with compressed air.

The specimen was then moved inside the installation room of the laboratory and was mounted on a specially designed installation fixture.

At low stress level, no surface cracks were observed. However, as the load increased, hair cracks were observed at the top surface of the horizontal member (unreinforced surface). At this point, a drop in the connection stiffness of about 23% was observed as shown in Figure (7). Due to the discontinuity of the composite laminates at the bottom faces of the horizontal member at the joint boundary (point of maximum moment on the horizontal member), similar cracks were developed at the bottom boundaries of the joint. The first local failure of the REPLARK laminate occurred at a load level of 29 kips (129 kN). The failure initiated at the lower left corner of the joint and was propagated upward along the boundary of the vertical member. Near the ultimate load, the tension crack in the composite laminate was extended vertically along the boundary of the joint. At this point, a sudden failure occurred due to a combination of concrete crushing and fracture of the all three top steel rebars of the horizontal member as shown in Figure (8).

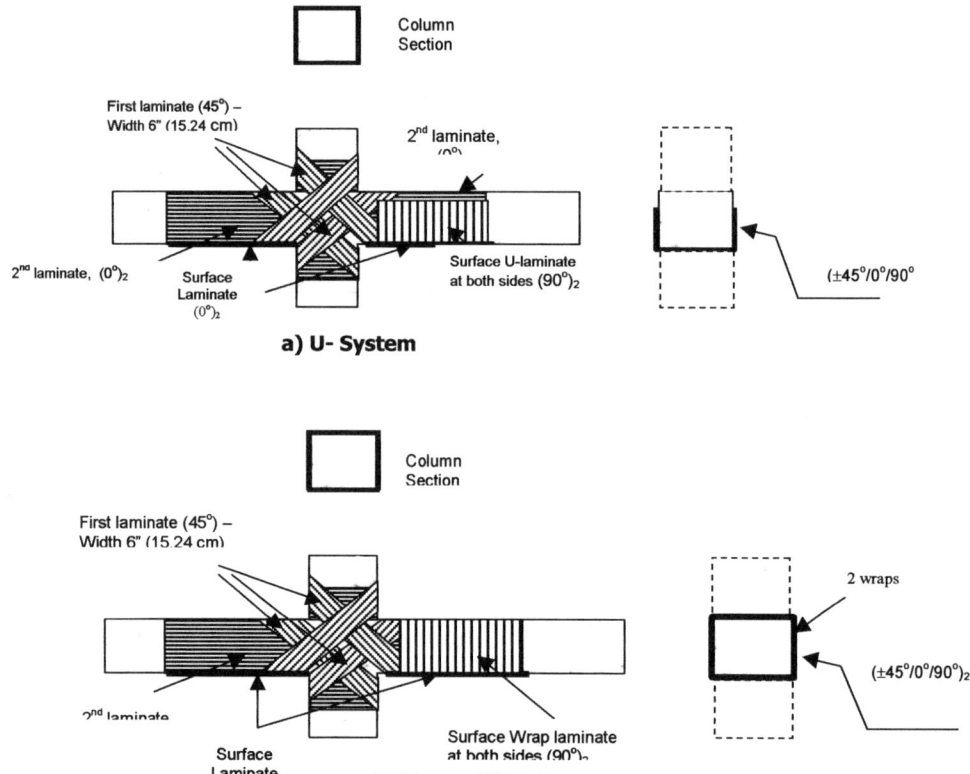

Figure (5): Common Laminates for U- and W- Systems

Test results indicated that the use of the composites not only succeed in restoring the ultimate capacity of the damaged specimen (P_u^{C1} = 21.83 kips [97 kN]) but also resulted in a strength gain of more than 34% (P_u^{RU} = 29.3 kips [130.32 kN]). In addition, an upgrade of 59% in the connection initial stiffness was achieved. However, and unlike the control specimen, both stiffness and strength degradation were observed for the repaired specimen RU.

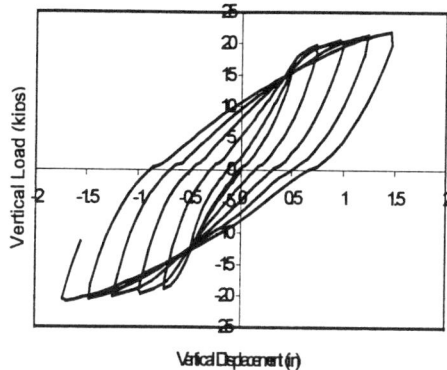

Figure (6): P/δ Hystrises for Control Specimen C1

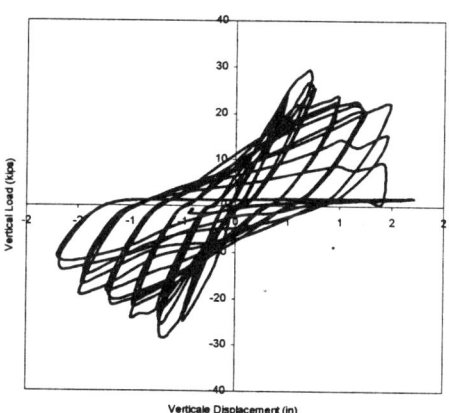

Figure (7): P/δ Hystrises for the Retrofitted Specimen **RU**

Retrofitted Specimen RTU: *Same* lamination schedule and details were used for this "undamaged" specimen. Both load-control and displacement-control regimes were used. The behavior of this specimen was similar to the repaired specimen RU. Initially the connection exhibited a high stiffens which lasted for the first four cycles after which an appreciable drop in the connection stiffness (about 79% of the initial stiffness) coupled with a slight strength degradation (about 12% of the initial connection strength) were observed as shown in Figure (9). Similar to the repaired specimen RU, the initial failure was initiated at the boundaries of the joint, and the local failure was initiated at the bottom surface of the horizontal member. As the load increased, the damage propagated vertically. Simultaneously, increase in both sizes and propagation of the horizontal member's top concrete surface was observed. The ultimate failure mode was a combination of a complete failure of the composite laminates at the joint boundary followed by a concrete crushing and repute of all three rebars at the left side and two bottom rebars of the horizontal member as shown in Figures (10). An increase in the connection stiffness up to 38% was achieved when composite overlays were used. In addition, a 38% increase in the displacement at failure was achieved as compared to the un-reinforced joint specimen. However, a slight increase in the strength resulted from applying the composites. This can be attributed to the lack of confinement, as well as the lack of composite reinforcement at the top of the horizontal member.

Figure (8): Fracture of Steel Reinforcements of Retrofitted Specimen **RU**.

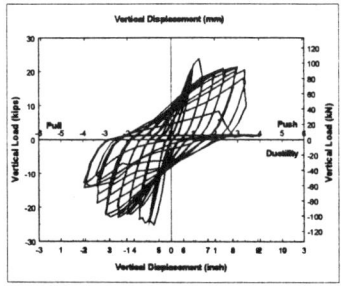

Figure (9): P/δ Hystrises for the Retrofitted Specimen **RTU**

Figure (10): Ultimate Failure of Retrofitted Specimen **RTU**

Repaired Specimen RW: After exposing the control specimen C2 to a sever cyclic loadings which resulted in major damage to both steel (beyond yield) and concrete, the joint was repaired using both epoxy injection and composite laminates. Same procedures which was applied to the repair of the first control specimen C1 was followed. This including removing all loose concrete, cleaning and applying epoxy mortar and epoxy injection techniques. For this specimen, a W-system was used in laminating the exterior surfaces of the specimen as shown in Figure (5-b). In this case, a laminate was wrapped around the cross-section of the horizontal members from the two sides of the vertical member in addition to the same lamination used to the U-system. At low stress level, the connection exhibited high stiffness as shown in Figure (11). For the first five cycles, the average axial stiffness was 50 kip/in (8.75 kN/mm) with an average rotational stiffness of 90,000 kip-in/rad (10,161 kN-m/rad). This can be translated in an increase of 167% as compared to the average stiffness of the control specimen C2 (axial stiffness = 18.75 kip/in (3.28 kN/mm). In addition, the connection exhibited high strength behavior in the first five cycles up to 35 kips (156 kN) as compared to an average strength of 20.35 kips (90.52 kN) for the control specimen C2. This is about 75% increase in the joint strength. After five cycles, the strength degradation was about 29% of the initial strength and was steady until the last three cycles where appreciable strength degradation took place as shown in Figure (11).

Similar to specimens RU and RTU, and due to the discontinuities of the fibers at both the top and bottom surface of the horizontal members, the failure was initiated at the maximum moment section of the extreme surfaces (top & bottom) of horizontal member adjacent to the connection where no composite reinforcement is provided. This crack was propagated through the thickness and was resisted mainly by the 0°-laminates bonded to the side surfaces of the horizontal member. The ultimate failure of specimen RW was a combination of laminate tensile failure as well as fracture of the steel rebars at both top and the bottom of the horizontal member.

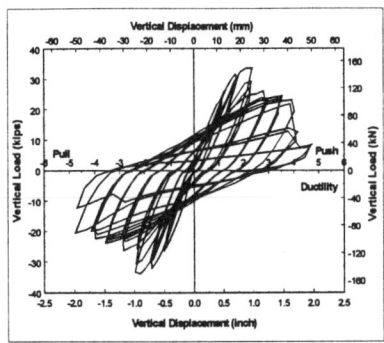

Figure (11): P/δ Hystrises for Repair Specimen **RW**

Repaired Specimen RTW: In this detail, "undamaged" connection specimen, with the same dimensions and lamination schedule as for the repaired specimen RW, was tested under both load- and displacement-control full reversal loading regimes. Similar to the RW specimen, this connection specimen exhibited high stiffness and strength at the first seven cycles, after which a slight degradation in strength coupled with appreciable stiffness degradation was observed as shown in Figures (12). At low stress level, the no surface cracks were observed. The direction of the initial crack was first diagonally and then moved vertically parallel to the side of the joint. The initial stiffness of the connection was 20% higher than the initial stiffness of specimen RW, and 60 kip/in (10.51 kN/mm) and 220% higher than the corresponding stiffness of the control specimen C2 (axial stiffness of C2 = 18.75 kip/in (3.28 kN/mm)). In addition, the connection exhibited high strength behavior in the first five cycles up to 35 kips (156 kN) as compared to an average strength of 20.35 kips (90.52 kN) for the control specimen C2.

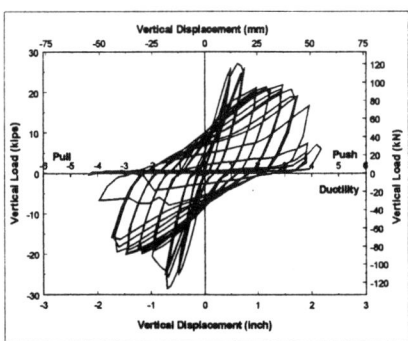

Figure (12): P/δ Hystrises for Retrofitted Specimen **RTW**

ANALYTICAL MODEL

Model Description: A computer code, based on moment-curvature analysis, was developed to analyze both control and retrofitted specimens [4]. For material constitutive relationships, the core concrete in the control specimen was considered confined by the steel ties, and Mander's model for concrete confinement [6] was employed. For retrofitted specimen, concrete was assumed confined by the composite jacket only, and Hosotani's Model for FRP-confined concrete [5] was utilized. In addition to defining the stress-strain characteristics of concrete, the stress-strain properties of steel reinforcement were also implemented. Steel was assumed to have three distinct major zones: a linear portion up to the yield, a yield plateau region, and a parabolic strain-hardening zone. In modeling, it was assumed that the role of the composite jacket was limited to confining action in order to increase both the strength and maximum compressive strain of concrete section, and hence enhance the member ductility. Jacket contribution to the member flexural rigidity was neglected.

In order to get the moment-curvature relationship for the member cross-section, laminar analysis procedure was performed. The section was divided into 100 slices, 5 each in the top and bottom cover and 90 slices in the core region. The assumption of plane section remains plane was made so that the assumption of a linear strain variation across the member cross-section was valid. The strain in each layer was calculated and henceforth, the stress and force resultant in each layer was computed. The equilibrium of internal forces was assumed through use of convergence criteria. Axial force convergence allowed both the moment resultant and the section curvature to be calculated.

The load-displacement response was obtained by modeling the test specimen as simply supported member as shown in Figure (13). For each part of the member, curvature distribution was assumed to be linear up to the first-yield of the extreme tensile steel. After yield, the curvature was assumed to be constant over the plastic hinge length. The moment of curvature distribution about the center point of the member determined the mid-span deflection. Details of load-displacement calculations are as follows:

(1) get the applied load, P, from:
$$P = \frac{2M}{L_o} \tag{5}$$
where M is the moment at the member critical section

(2) For $\Phi < \Phi_y$:
$$\Delta = \phi L_o (L_o / 3 + Z / 2) \tag{6}$$
where Φ_y is the first-yield curvature, and L_o and Z are as defined in Figure (13).

(3) For $\Phi = \Phi_y$:
$$\Delta_y = \phi_y L_o (L_o / 3 + Z / 2) \tag{7}$$
where Δ_y is the first-yield displacement

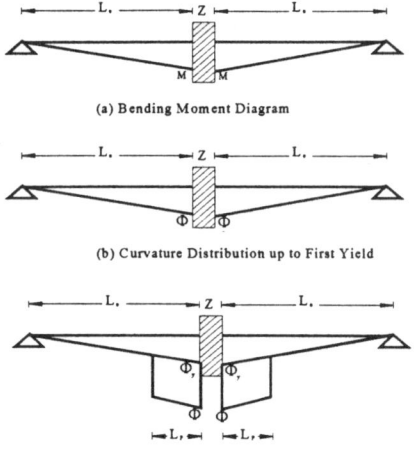

Figure (13): Specimen Modeled as Simply Supported Member.

(4) For $\Phi > \Phi_y$:

$$\Delta = \Delta_y + \Delta_p \tag{8}$$

where Δ_p is the plastic displacement, and is given by:

$$\Delta_p = (\phi - \phi_y)L_p(L_p + Z) \tag{9}$$

where L_p is the plastic hinge length, and is calculated according to Priestley et al. [4] by:

$$L_p = 0.08L_o + 0.15 f_y d_{bl} \geq 0.3 f_y d_{bl} \tag{10}$$

where f_y and d_{bl} are the yield strength and bar diameter of the longitudinal steel, respectively.

THEORETICAL RESULTS

Theoretical results in terms of load-displacement envelopes for both control and retrofitted specimens are as shown in Figures (14-a) and (14-b), respectively. For control specimen, it was shown that load increased with displacement increase up to failure which was due to low cycle fatigue fracture of tensile steel. The same trend was observed for retrofitted specimen. Comparison between experimental and theoretical results showed quite well agreement in case of control specimen and good agreement for retrofitted specimen. Summary of theoretical and experimental results is shown in Table (1).

Table (1): Experimental and Theoretical Results for Test Specimens

Test Specimen	Experimental Results			Theoretical Results			Comparison		
	P_u (kips)	Δ_u (in)	μ_Δ	P_u (kips)	Δ_u (in)	μ_Δ	$P_{u\text{-exp}}/P_{u\text{-th}}$	$\Delta_{u\text{-exp}}/\Delta_{u\text{-th}}$	$\mu_{\Delta\text{-exp}}/\mu_{\Delta\text{-th}}$
Control	19.55	1.87	5.5	22.06	2.36	7.0	0.89	0.79	0.79
Retrofitted	20.77	1.69	5.1	23.79	1.62	4.9	0.87	1.05	1.05

Note: P_u: Maximum applied Load;
Δ_u: Ultimate vertical displacement;
μ_Δ: Ultimate displacement ductility

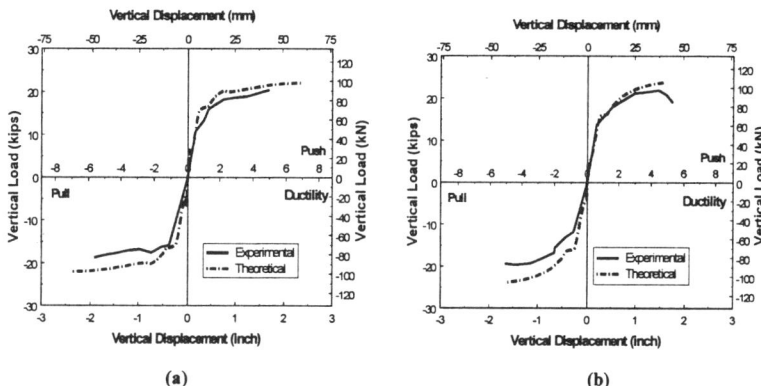

(a)   (b)
Figure (14): Load-Displacement Response for: a) Control Specimen **C2**
b) Retrofitted Specimen **RTW**

CONCLUSIONS

Experimental and theoretical results on seismic response of reinforced concrete joints repaired and retrofitted with composite laminates were presented in this paper. For the U- repair system, test results indicated that the use of the carbon/epoxy composite laminates not only succeed in restoring the ultimate capacity of the damaged specimen (P_u^{CI}= 21.83 kips [97 kN]) but also resulted in a strength gain of more than 34% (P_u^{RU}= 29.3 kips [130.32 kN]). In addition, an upgrade of 59% in the connection initial stiffness was achieved. However, and unlike the control specimen, both stiffness and strength degradation were observed for the repaired specimen RU. This can be attributed to the fact that the steel reinforcements of the repaired connection yielded before conducting the test. In addition, and due to the use of the epoxy injection repair, both stiffness and strength degradation are expected when progressive brittle failure occurred to the injected epoxy. Also, due to the fact that this repair detail did not cover all the faces of the concrete section (U-schedule), the top pre-cracked concrete surfaces (with already yielded steel) were exposed to very high tensile and compression stresses. For the W system, an increase of the initial stiffness up to 167%, as compared to the average stiffness of the control specimen, was achieved. In addition, a 75% increase in the initial joint strength was achieved. Comparison between experimental and theoretical results for the control joint and joint retrofitted with the W-system (RTW) showed quite well agreement.

ACKNOWLEDGMENTS

This study was sponsored by Mitsubashi Chemicals. The valuable technical remarks of Mr. Yagi and Mr. Ganjehlou of Mitsubashi Chemicals, and the support of Mr. Kano of Sumitomo Corporation are highly appreciated. The contributions of Wallock & Maggio Construction Company in supplying and performing the repair and the epoxy injection is appreciated. The author would like also to acknowledge the remarkable effort of Mr. James Kiech of CSUF in fabricating the test fixture and operating the full-scale tests.

REFERENCES

1. Advanced Seismic Repair & Rehabilitation Structural System, Proceedings, CSUF State-Of-The-Art Conference on Repair and Rehabilitation of Structures, Editor: A. Mosallam, Fullerton, CA, November 1998, 9201p.
2. Mosallam, A.S., Chakrabarti, P.R. and Lau, E.K. (1999) *"Concrete Concertinos"*, Civil Engineering Magazine, ASCE, Vol. 69, No. 1, January 1999, pp. 42-44.
3. Issa, M., Al-Chaar, G., Islam, M., Leslie, M., Abdalla, H., and Valle, C. (2000). *"Evaluation of CFRP Composites for Sesmic Retrofit of Reinforced Concrete Structures,"* Proceedings. Edt. A. Mosallam, SRRS2 Conference, Fullerton, California, March.
4. Haroun, M.A., Elsanadedy, H.M. *"Prediction of Cyclic Performance of Composite-Jacketed Squat Reinforced Concrete Bridge Columns,"* Proceedings, Composites in Transportation Conference (AUCN-2), Sydney, Australian, February 14-16.
5. Hosotani, M., Kawashima, K., and Hoshikuma, Jun-ichi, *"A Stress-Strain Model for Concrete Cylinders Confined by Carbon Fiber Sheets (in Japanese),"* Report No. TIT/EERG 98-2, Tokyo Institute of Technology, Tokyo, Japan, 1998, 55pp.
6. Mander, J.B., Priestley, M.J.N., and Park, R., *"Theoretical Stress-Strain Model for Confined Concrete,"* Journal of the Structural Division, ASCE, Vol. 114, No. 8, August 1988, pp. 1804-1826.

Seismic Retrofit of Reinforced Concrete Members with CFRP Composites

M. A. ISSA, M. S. ISLAM, M. LESLIE, H. ABDALLA and C. DO VALLE

ABSTRACT

The structural behavior of beams, beam-column joints, and cylinders confined by means of carbon fiber reinforced polymer (CFRP) wrapping is presented in this paper. At present, most of the old structures located in high seismic zones are either deteriorated or structurally deficient. Solutions for the upgrade of these structures using economical and reliable techniques are of great interest. This research was aimed at studying the application of CFRP composites for retrofitting or strengthening of these structures.

A total of 6 beams, 2 beam-column joints and 6 cylinders were cast, of which half were wrapped with CFRP sheets, and tested. Test results indicated that the externally strengthened beam displayed an increase in ultimate load carrying capacity by as much as 30%. It was observed that wrapped joints could resist more loading cycles than unwrapped joints. The observed ductility for the wrapped specimen was 30% higher than that of the unwrapped specimen although one thin layer of CFRP was applied. The compressive strengths of cylinders wrapped with one layer were 50% higher than those of the unwrapped cylinders. A significant increase in stiffness was observed in the wrapped cylinders.

INTRODUCTION

Concrete bridges and other reinforced concrete structures are constantly deteriorating and are in need of major maintenance or total repair. Recent earthquakes around the world caused major damage to many structures. As a result, the social demand for seismic diagnosis and retrofit of existing concrete structures became stronger as it accelerated studies aimed to establish technology for strengthening existing buildings and bridges.

For many years, the application of epoxy bonded steel plates on concrete flexural members was performed as a repair technique worldwide where evidence of damage existed. Despite the promise of a cheap and effective solution, some major problems were discovered with bonding, corrosion and placement of the steel plates. For these reasons and many others, steel plate bonding as a repair procedure has never

---

Mohsen A. Issa, Md. S. Islam, Michael Leslie, Hiba Abdalla and Cyro do Valle, Department of Civil and Materials Engineering, University of Illinois at Chicago, 842 W. Taylor Street, Chicago, IL 60607.

gained great acceptance. In recent years, new materials and technologies have been developed. The marine and aerospace industries have developed plastics and adhesives with characteristics very dissimilar to most conventional materials. These materials have advantages that are typically demonstrated in their generic general properties, such as lower weight, high tensile strength, modulus, corrosion, fatigue resistance, etc. These new technologies should not be accepted based upon their relative promise of innovation. Further research must be performed into each and every application of fiber reinforced polymers (FRP), before it can be accepted as a structural material.

The first use of FRPs in structural rehabilitation came with the introduction of high strength FRP plates. Next it was used for externally bonding thin FRP sheets and FRP fabric materials to concrete. The use of fabrics and sheets easily overcame problems demonstrated in both plates and channels. The material forms a skin around the concrete member with the surface fully bonded and forming a shape that confines the member. Fiber reinforced polymer sheets are superior to steel plates in that they offer high resistance to electrochemical corrosion, a high strength to weight ratio, ease of handling, the ability to be formed in any shape, fatigue resistance, and their availability in any length. Carbon fiber reinforced polymer (CFRP) sheets have been applied to concrete columns, piers, beams, girders, slabs and walls for retrofit. Retrofitting of concrete structures shall be performed with the aim to ensure safety during earthquakes as well as to prevent fatal damage affecting human life and degradation of structural function affecting the life and work productivity of inhabitants. In order for the application of continuous fiber sheets to become a promising seismic retrofitting technology, it is important to clearly understand the existing problems and determine the future direction of structural evaluation. The main objective of this research program was to investigate the effects of the application of CFRP sheets on stiffness, structural performance, ductility, and mode of failure of reinforced concrete structural members.

LITERATURE REVIEW

The use of externally bonded fiber reinforced polymers (FRP) has gained widespread acceptance as an excellent method for the strengthening, retrofitting and upgrading of existing concrete structures. This strengthening method has many advantages over the traditional techniques due to the high strength to weight ratio, ease of installation on site, and the improved durability and corrosion resistance of the composite material [1]. The efficiency of externally bonded FRP reinforcement in terms of structural performance and ease of application has been demonstrated [2]. As long as there is a perfect bond between the carbon fiber sheet and concrete, the traditional flexural theory for RC members is applicable to calculate the contribution of the carbon fiber sheet. The contribution of multi-layers of sheets may not be deduced from that of a single layer, i.e., the strength does not increase linearly [3].

Catbus [4] investigated the performance of reinforced concrete beams retrofitted using CFRPs. It was found that shear failure becomes the primary mode of failure if CFRPs are applied to the tension face. This shows a shift in failure mode from flexural (ductile) to shear (brittle). The preferred mode of failure for the composite system is by crushing of the concrete in compression. For this reason, CFRP should be placed along the sides of beams to provide for additional shear reinforcement to achieve the preferred mode of failure. Grace et al. [5] investigated the use of carbon fiber reinforced polymer sheets to strengthen positive and negative moment regions of continuous beams. It was concluded that the use of FRP laminates to strengthen continuous beams is effective in reducing deflections and increasing their load carrying capacity. Furthermore, beams strengthened with FRP laminates exhibit smaller and better distributed cracks. Taly and GangaRao [6] concluded that design using FRPs should be based on Load Resistance Factor Design and focus on compression failure of concrete and not yielding of the steel in order to avoid brittle failure of the FRP. Shear strength of FRPs depends primarily on the properties of the resin. The externally bonded reinforcement could be used to enhance the shear capacity of the beams in positive and negative moment regions by as much as 22 to 135%. The CFRP contribution is enhanced to a large degree for beams without stirrups than for beams with adequate steel shear reinforcement [7].

FRPs provide effective confinement for columns subject to lateral drift, up to $0.5\ f'c\ A_g$. The number of layers of FRP is dependent on the underlying assumptions, especially those with respect to plastic hinge formation [8]. Shmoldas et al. [9] investigated the retrofit of a rectangular column using carbon fiber for the Arroy Sero Bridge in Pasadena, California. A full scale test was performed to assess the effectiveness of confinement in the column in strong and weak axis bending. The column has a high, 3 to 1, aspect ratio. The results of the test showed no increase in ductility for weak axis bending, but a doubling of ductility for the strong axis as compared to the built in place. Zhang et al. [10] examined the seismic performance of existing RC columns retrofitted with carbon fiber sheets. Quantity, retrofitting types and kinds of continuous fiber sheets were adopted as the experimental variables. The failure mode of the columns changed from shear to flexural failure. However, if the ply of continuous fiber sheet is not enough, the columns collapse due to crushing of concrete just after yielding of longitudinal rebar and the ductility is poor.

Application of CFRP improved the flexural and shear capacity of structural members. A normal column that was shear-strengthened by carbon fiber sheet changed the failure mode from shear failure to flexural failure with improved ductility, as the amount of shear reinforcement was increased [11]. Pantelides et al. [12] found that the CFRP composite strengthened the cap beam-column joints effectively for an increase in shear stresses of 35 percent. The bent retrofitted with CFRP composite reached a system displacement ductility of 6.3 as compared to the bent in the as-is condition, which reached a ductility of 2.8. The peak lateral load capacity was increased by 16%.

MATERIALS, SPECIMEN CONFIGURATION AND PREPARATION

Two concrete mixes were prepared. Mixture 1 consisted of Type I Portland cement, fly ash, regular river sand with a maximum aggregate size of $^3/_{16}$", and coarse aggregates of ¾" maximum size. Mixture 2 consisted of Type I Portland cement, regular river sand with a maximum aggregate size of $^3/_{16}$", coarse aggregates of ⅜" maximum size and superplasticizer. Both mixture proportions are shown in Table I.

Mixture 1 was used for beams and cylinders, while Mixture 2 was used for the beam-column joint. Concrete of Mixture 1 and Mixture 2 had 28-day compressive strengths of 6,000 psi and 6,500 psi, respectively. The steel bars used had a yield strength of 56 ksi. Table II gives an overview of the main properties of the CFRP wrapping sheet used in this study. Six beams, two beam-column joints (cruciform shaped), and six cylinders were fabricated and tested. The 4.5" x 9" x 96" beam was reinforced with 2#5 bottom longitudinal reinforcing steel and 2#3 top reinforcement. Closed stirrups of 0.25" diameter were used as shear reinforcement. One beam was wrapped on the tension face, one beam was fully wrapped with one layer of CFRP and one was fully wrapped with two layers of CFRP. The other three beams were unwrapped. Each joint specimen consisted of a 5" x 7.75" beam and a 5" x 8" column, with 22" and 30" length on each side of the joint, respectively. The beam for the joint was reinforced with 4#3 bottom longitudinal reinforcing steel with a ¼" discontinuity at the center of the joint, and 2#3 top reinforcement. Closed shape $^3/_{16}$" diameter stirrups were used with a spacing of 3" as shear reinforcement. The column of the joint was reinforced with 4#3 bars with rectangular ties of $^3/_{16}$" diameter bar and a spacing of 5". One specimen was completely wrapped with one thin layer of CFRP sheet and the other was unwrapped. In addition, three cylinders were circularly wrapped with one layer of CFRP sheet, providing sufficient overlap (3 in.).

TABLE I. MIXTURE PROPORTIONS OF CONCRETE

Ingredients	Mix proportions for 1 yd^3	
	Mix # 1	Mix # 2
Cement, lb	548	575
Fly ash, lb	97	--
Water, lb	323	316
Coarse aggregate, lb	1485	1772
Fine aggregate, lb	1388	1176
RB 3000 superplasticizer, fl oz	--	23

TABLE II. PROPERTIES OF CARBON FIBER SHEET

Properties	Values
Tensile strength	530 ksi
Tensile modulus	33.5 x 10^6 psi
Density	0.064 lb./in^3
Elongation at break	1.4%
Elastic recovery	100%
Filaments/strand	12,000
Sheet thickness	0.026 in

Corners of the rectangular specimens were smoothed with a curvature to reduce stress concentrations on the CFRP sheets. After 28 days from casting time, the CFRP sheets were applied to the specimens. The surfaces of the beams, joint, and cylinders were prepared by sandblasting until the coarse aggregate was slightly exposed and the dust was removed from the concrete surface using air pressure. A base of epoxy resin was mixed with a hardener using a hand-powered mixer. Epoxy was applied to the concrete surface and CFRP sheet simultaneously. One side of the CFRP sheet was placed on the specimen and pressed while a second person applied the remainder of the sheet forcing it in one continuous movement. To remove any entrapped air, the surface of the CFRP sheet was pressed with a small hand roller. Afterwards, all specimens were allowed to cure at laboratory conditions until testing.

TESTING PROCEDURE

Strain gages and rosettes were mounted at critical locations to measure the strain. The layout of strain gages and rosettes for a typical beam-column joint is shown in Figure 1. LVDT's were installed to measure the displacements. Beam specimens were tested using four point loading as shown in Figure 2. Load was applied in a displacement control mode at a rate of 0.01 in./min. In beam-column joint testing, the loading pattern was a cyclic type with alternating displacement reversals (Figure 3). The beam was connected to an actuator and was subjected to push and pull, while the column was simultaneously subjected to a constant axial load of 50 kips which is about 20% of the ultimate compressive load of the column. The test setup for a typical beam-column joint is shown in Figure 4 (a). Strain gages were also mounted on cylinder specimens in both transverse and longitudinal directions. Load, deflection, and strain readings were recorded via a data acquisition system and the development of cracks and mode of failure were monitored throughout the test of all specimens.

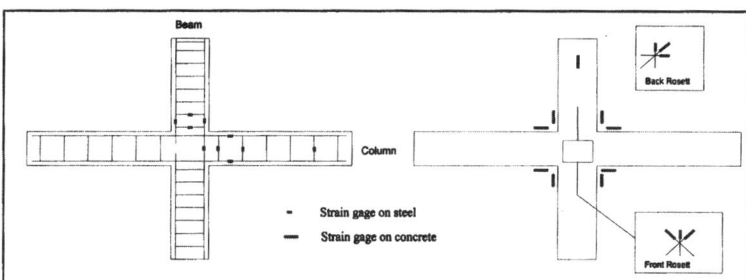

Figure 1. Instrumentation on a typical beam-column joint.

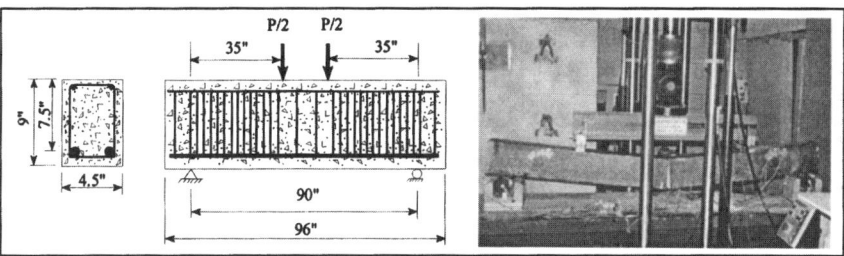

Figure 2. Beam configuration and testing setup.

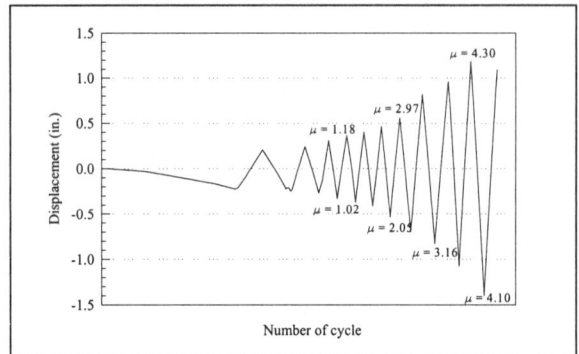

Figure 3. Loading cycles for typical beam-column joint testing.

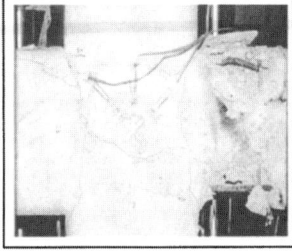
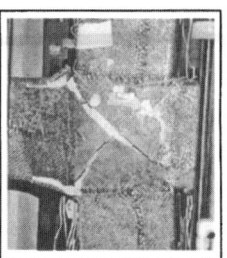

(a) Typical test setup for beam-column joint

(b) Failure mode of unwrapped joint

(c) Failure mode of wrapped joint

Figure 4. Test setup and failure mode of wrapped beam-column joint.

RESULTS AND DISCUSSION

A total of 6 beams, 2 beam–column joints and 6 cylinders were tested. The observed results were analyzed and compared. The load-deflection response of CFRP wrapped and unwrapped beams are shown in Figure 5. It was observed that CFRP wrapping can increase the flexural capacity by as much as 30%, however it does not add to the stiffness of beams. It can be seen from Figure 5 that when the tension face wrapped beam reached its ultimate strength, a sudden drop of load was experienced and afterwards, the load-displacement curve followed the trend of unwrapped beam. On the other hand, for fully wrapped beams, although a fall of load was observed, they continued to carry higher loads than those of unwrapped or tension face wrapped beams. This is because of the contribution of the CFRP sheets in the web of the beams. In all wrapped beams, failure occurred due to tensile cracking of the laminate following the crushing of the concrete under the loading points.

The load-displacement hysteresis curves for unwrapped and wrapped beam-column joints are shown in Figures 6 and 7, respectively. The shearing strength of the wrapped joint was 13% higher than that of the unwrapped joint although a very thin layer of CFRP was used. The ultimate failure in the unwrapped joint specimen was due to shearing of the column, while in the case of the wrapped joint, the ultimate failure was within the joint at an angle of 45° as shown in Figures 4(b) and 4(c). It was observed that wrapped specimens could resist more loading cycles than unwrapped specimens. It was also noticed that the wrapped specimen could undergo 86% more displacement than the unwrapped specimen as depicted in Figures 6 and 7.

The comparison of wrapped and unwrapped cylinder test results is shown in Figure 8. The confinement and stiffness due to wrapping with CFRP sheets was clearly manifested. The ultimate load for the wrapped cylinder was 50% higher than that of the unwrapped cylinder although a very thin layer of CFRP was applied.

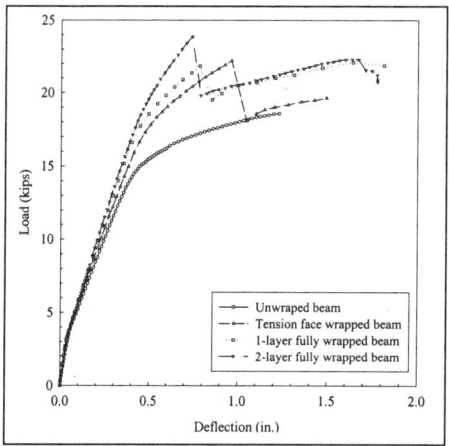

Figure 5. Load vs. deflection of wrapped and unwrapped beams.

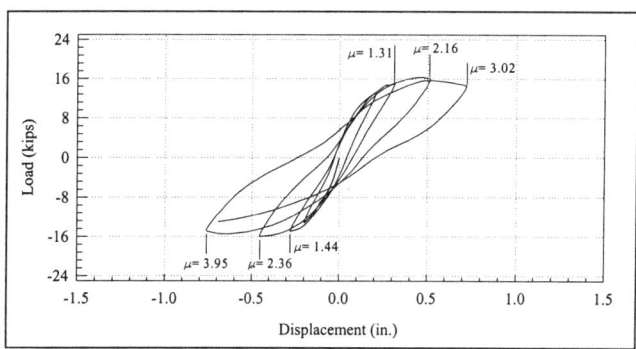

Figure 6. Load vs. displacement of unwrapped beam-column joint.

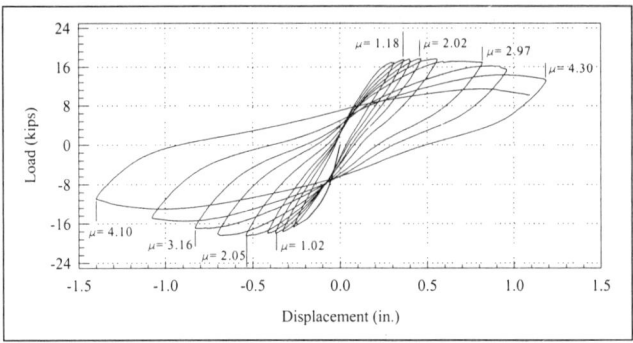

Figure 7. Load vs. displacement of CFRP wrapped beam-column joint.

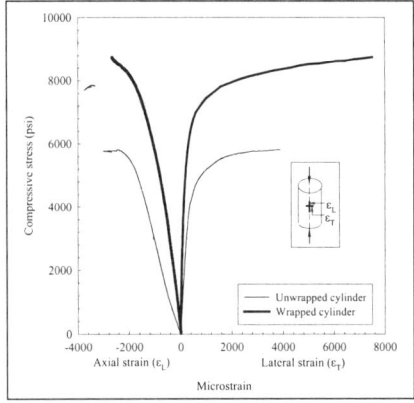

Figure 8. Typical compressive strength vs. axial and lateral strain of wrapped and unwrapped cylinders.

CONCLUSIONS

The test results provided strong evidence that retrofit using CFRP composites is a viable option for improving the seismic performance of reinforced concrete structures. An extensive experimental program is underway to investigate the durability and structural behavior of beams, columns, slabs and beam-column joints wrapped with CFRP sheets and strips and subjected to severe environment, and sustained and fatigue loading. Based on the results of the ongoing project, the following conclusions can be drawn:

1. Externally strengthened beams demonstrated an increase in ultimate load carrying capacity by as much as 30%.
2. The compressive strengths of the wrapped cylinders were 50 percent higher than those of the unwrapped cylinders. The stiffness of wrapped cylinders

is higher than that of unwrapped cylinders. Wrapped cylinders experienced lateral strain two times higher than that of unwrapped cylinders.
3. The shearing strength of the beam-column joint wrapped with one thin layer of CFRP was 13% higher than that of the unwrapped joint.
4. The wrapped joints resisted more loading cycles than unwrapped joints. The observed ductility of the specimen wrapped with one thin layer of CFRP was 4.3, while for the unwrapped specimen the ductility was 3.0.

REFERENCES

1. Meier, U. 1997. "Post Strengthening by Continuous Fiber Laminates in Europe, Non-Metalic (FRP) reinforcement for Concrete Structures," Proceedings of the Third International Symposium, 1:42-56.
2. Nanni, A., Ed. 1993. "Fiber-Reinforced Plastic Reinforcement for Concrete Structures: Properties and Application," Developments in Civil Engineering, Elsevier, 42:450.
3. Maruyama, K., Ueda, T., Hoshijima, T., and Uemura, M. 1999. "Japan Society of Civil Engineers Activity on Continuous Fiber Sheet for Retrofit of Concrete Structures," Proceedings of the Fourth International Symposium on Fiber Reinforced Polymer Reinforcement for Reinforced Concrete Structures, Eds. Charles et. al., ACI International, pp. 151-157.
4. Catbas, K. H. 1997. "Performance of Beams Externally Reinforced with Carbon Fiber Reinforced Plastic Laminates," Masters Thesis, Department of Civil and Environmental Engineering, University of Cincinnati, pp. 92.
5. Grace, N.F., Soliman, A.K., Abdel-Sayed, G., and Saleh, K.R. 1999. "Strengthening of Continuous Beams Using Fiber Reinforced Polymer Laminates," Proceedings of the Fourth International Symposium on Fiber Reinforced Polymer Reinforcement for Reinforced Concrete Structures, Eds. Charles et. al., ACI International, SP 188-57, pp. 647-657.
6. Taly, N., and GangaRao, H. V. S. 1999. "Guidelines for Design of Concrete Structures Reinforced with FRP Materials," 44th International SAMPE Symposium, May 23-27, 1999, pp. 1689-1696.
7. Khalifa, A., Tumialan, G., Nanni, A., and Belarbi, A. 1999. "Shear Strengthening of Continuous Reinforced Concrete Beams Using Externally Bonded Carbon Fiber Reinforced Polymer Sheets," Proceedings of the Fourth International Symposium on Fiber Reinforced Polymer Reinforcement for Reinforced Concrete Structures, Eds. Charles et. al., ACI International, SP 188-84, pp. 995-1008.
8. Nosho, K. J. 1996. "Retrofit of Rectangular Reinforced Concrete Columns using Carbon Fiber," Masters Thesis, Department of Civil Engineering, University of Washington, pp. 194.
9. Shmoldas, A., Schleifer, G., Seible, F., and Innamorato, D. 1997. "Carbon Fiber Retrofit of the Arroyo Seco Spandrel Column," Report No. SSRP-97/13, Division of Structural Engineering, University of California, San Diego.
10. Zhang, A., Yamakawa, T., Zhong, P., and Oka, T. 1999. "Experimental Study on Seismic Performance of Reinforced Concrete Columns Retrofitted with Composite-Materials Jackets," Proceedings of the Fourth International Symposium on Fiber Reinforced Polymer Reinforcement for Reinforced Concrete Structures, Eds. Charles et. al., ACI International, SP 188-24, pp. 269-278.
11. Matsuzaki, Y., Nakano, K., Fujii, S., and Fukuyama, H. 1999. "Japanese State of the Art on Seismic Retrofit by Fiber Wrapping for Building Structures: Research," Proceedings of the Fourth International Symposium on Fiber Reinforced Polymer Reinforcement for Reinforced Concrete Structures, Eds. Charles et. al., ACI International, SP 188-75, pp. 879-893.
12. Pantelides, C.P., Gergely, J., Reaveley, L.D., and Volnyy, V.A. 1999. "Retrofit of Reinforced Concrete Bridges with Carbon Fiber Reinforced Polymer Composites," Proceedings of the Fourth International Symposium on Fiber Reinforced Polymer Reinforcement for Reinforced Concrete Structures, Eds. Charles et. al., ACI International, SP 188-40, pp. 441-453.

Seismic Retrofit of Reinforced Concrete Columns Using FRP Composite Laminates

M. HAROUN, M. FENG, M. YOUSSEF and A. MOSALLAM

ABSTRACT

The paper presents summary of the results of a comprehensive analytical and large-scale experimental investigation on the behavior of reinforced concrete columns with Fiber Reinforced Polymer (FRP) Composite Jackets. In the experimental phase, a total of six half-scale reinforced concrete columns, three with and three without lap splices, were built based on the old seismic design code to represent many existing old bridge columns in California. Two types of jackets were used in this study, one a wet-layup carbon/epoxy laminates, and the other is a machine-wound carbon/epoxy jacket. Two control "un-jacketed" columns were also tested to provide information on the ultimate behavior of un-retrofitted columns and to compare this behavior with the FRP jacketed specimens. Cyclic loading tests have demonstrated that the performance of such existing columns can be improved dramatically due to the enhancement of concrete confinement provided by these advanced composite jackets. The paper also describes a summary of an on-going comprehensive test program on the confinement behavior of reinforced and un-reinforced columns with composite jackets. The project is joint effort between University of California, Irvine, and California State University, Fullerton.

INTRODUCTION

Recent earthquakes (e.g. North Ridge, Kobe, Turkey, etc.) have indicated that bridges designed and built according to older seismic design codes are vulnerable to catastrophic collapse resulted from failure of the super structure, specifically the concrete columns. Elastic design principles used in earlier design methodology resulted in large structural members that created a false sense of security with designers which led to lack of concentration on detailing. As a result, concrete

---

M. Haroun, M. Feng, and M. Youssef, Civil and Environmental Engineering Department, University of California, Irvine, Irvine, California 92697-2175, USA
A. Mosallam, Division of Engineering, California State University, Fullerton, Fullerton, CA 92834, USA

columns in many existing bridges and buildings have the following potential problems i) insufficient lap-splice length which leads to inadequate flexural capacity, ii) insufficient shear strength, and iii) insufficient ductility performance due to inadequate transverse confinement. Another problem in concrete columns is the lack of lateral confinement and the associated lower energy absorption capacity. Many older reinforced concrete columns, especially those designed mainly to resist gravity loads contain very little lateral confinement reinforcements in the form of steel ties or hoops. Such level of confinement is inadequate to provide the desired seismic resistance by the structure.

For the past few years, steel jacketing has been proven as an effective technology for alleviating the problems and for enhancing the seismic performance of old bridge columns. Fiber composite systems, however, have gained wide recognition as strengthening measures to increase load carrying capacity in structural members. As a result these materials have shown great potential in becoming an attractive alternative as compared to steel jackets. Test results indicated that the use of composites, in particular those manufactured by the pultrusion process using carbon, glass, or aramid fibers, have resulted in high strength and excellent resistance against harsh environments. In addition, high strength fiber composite increases the ductility capacity and prevents brittle catastrophic structural failures. These characteristics make it very attractive as a retrofit system that can be used to improve seismic performance in seismic prone areas

In order to examine their seismic performance, experimental study was conducted at UCI on bridge columns retrofitted with two types of carbon fiber based composite materials made by Mitsubishi Chemical Corp. Two of the problems mentioned above with the old bridge columns were reflected in the design of six testing model columns: three with and three without lap splices. The columns with lap-splices were tested to evaluate the effectiveness of the composite jackets on enhancement of concrete confinement, while the columns with continuous longitudinal steel were tested for shear failure in a fixed-fixed condition.

REINFORCED CONCRETE COLUMN DETAILS

Type I Columns - with Lap Splices: Three 1:2 scale columns were built with a lap-splice. Each one has 20 #6 (dia = 1.905 cm) as the main longitudinal reinforcements with #2 (dia = 0.635 cm) spaced at 5 in (12.7 cm). The reinforcements are grade 40, as most existing columns with lap-splices have grade 40 reinforcement in the field. The footing are reinforced with grade 60 steel. The columns have a circular cross-section with a diameter of 24 in (60.96 cm). Other details are as shown in Figure 1 and Table 1. These columns were tested in single bending simulating a fixed-free condition with a moment arm of 144 in (3.66 m).

Type II Columns - without Lap Splices: Three 1:2 scale columns were built with continuous longitudinal reinforcements. The reinforcement and section details are shown in Figure 1 and Table 1, similar to the lap-splice columns, except that these columns have a height of 96 in (2.44 m) and were tested in double bending to simulate a fixed-fixed condition.

FRP RETROFIT DETAILS

Two types of composite material jackets were used: one is a carbon fiber sheet (brand name Replark30) applied by hands as shown in Figure 2a, and the other is a carbon fiber strand (henceforth referred to as CF) applied by a robot as shown in Figure 2b. One out of the three columns of each type was tested without a jacket (henceforth referred to as "as built" column), the second was retrofitted with Replark and the third was retrofitted with CF. Type I columns were retrofitted to enhance the confinement at the lap splice regions in order to improve the flexural capacity, while Type II columns were retrofitted to enhance the confinement in order to improve the shear capacity.

Table 1: Data for Test Columns

Diameter of section	24.00 in (60.96 cm)
Concrete strength	5.00 ksi (34.5 MPa)
Steel strength	43.41 ksi (299.3 MPa)
No. of main longitudinal bars	20
Main longitudinal bar size	6 (dia. = 1.905 cm)
Clear cover to confinement steel	0.75 in (1.905 cm)
Confinement steel bar size	2 (dia. = 0.635 cm)
Yield strength of confinement steel	30.5 ksi (210.3 MPa)
Spacing of confinement steel	5.00 in (12.7 cm)

DESCRIPTION OF THE EXPERIMENTAL PROGRAM

Test Procedure: The test setup was designed to subject the columns to a constant axial compressive load and cyclic horizontal loads. As shown in Figure 3(a), each of Type I columns was subjected to horizontal loading applied at the top of the column directly by an actuator installed on the strong wall to simulate the fixed-free condition.

As shown in Figure 3(b), each of Type II columns was subjected to horizontal loading applied through a test rig specially designed to restrain the rotation at the top of the column using a pantograph arrangement to simulate the fixed-fixed condition.

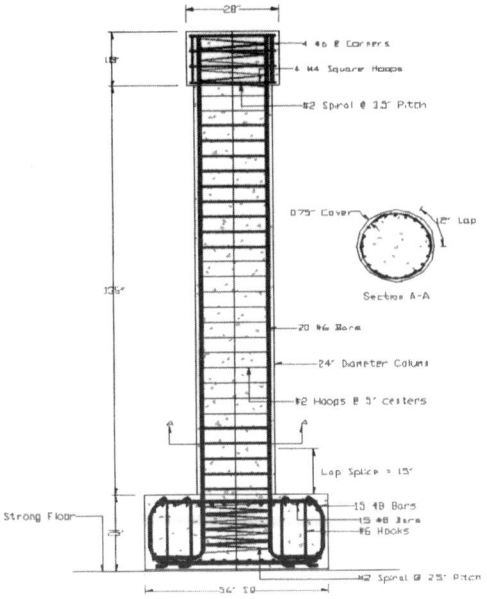

Figure 1. Reinforcement Details for Circular Lap Splice Columns

Figure 2. Reinforced Concrete Column Specimens with Carbon/Epoxy and E-glass/Epoxy

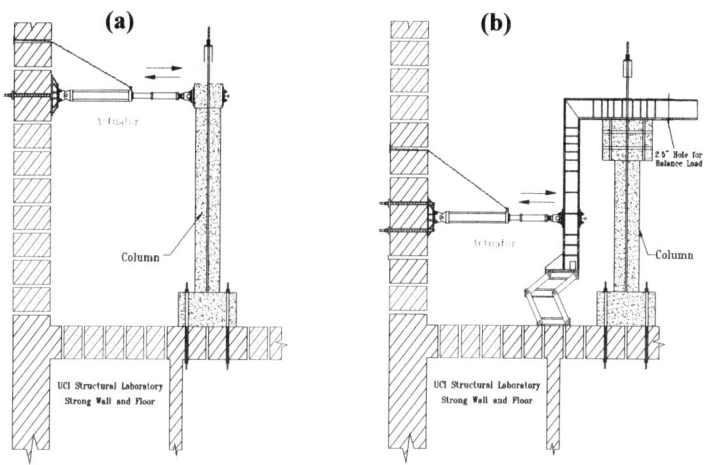

Figure 3: Test Setup: (a) Type I Column - Single Bending
(b) Type II Column - Double Bending

An axial of 145 kips (645 kN) was applied to the column by post-tensioning two steel rods with a hydraulic jack at the top of the column. The ratio of the applied axial load to the column axial load capacity is among the range of those typically used in multi-column bent bridges in California. The horizontal loading regime used in the tests (Table 2) is based on the guidelines of the California Department of Transportation (Caltrans). The initial loading cycles were controlled by the peak force until the column developed the calculated capacity corresponding to the first yield of longitudinal steel, H_y. Then, the test was stopped and the yield displacement was determined from:

Table 2: Horizontal Loading Regime

Load	Displacement	No. of Cycles
$0.25H_y$		3
$0.50H_y$		3
$0.75H_y$		3
H_y		3
	Δ_y	3
	$1.5\Delta_y$	3
	$2\Delta_y$	3
	$3\Delta_y$	3
	⋮	⋮

$$\Delta_y = \frac{H_i}{H_y} \Delta_1 \qquad (1)$$

Where Δ_1 was average of the measured peak displacements corresponding to the first yield lateral load capacity, H_y, in the push and pull directions. The ideal flexural lateral load capacity, H_i, was computed based on an extreme concrete compressive strain of 0.005 (0.004 for the unjacketed columns) in the column critical region and measured material properties. After the column developed the first yield capacity, loading cycles were controlled by the peak displacement.

EXPERIMENTAL RESULTS

For Type I columns, the as built column failed at a displacement ductility of 2, whereas the jacketed columns performed much better with displacement ductility of greater than 6. The envelope of the hysteresis obtained is compared in Figure 4. The maximum lateral loads obtained are tabulated in Table 3.

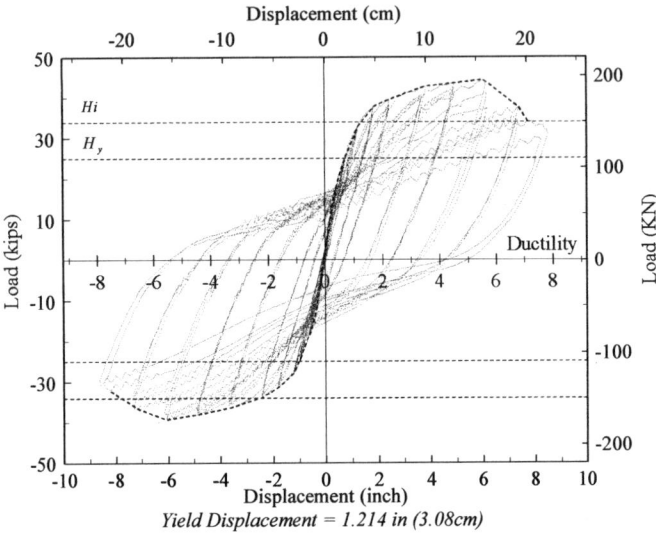

Figure 4. Hysteresis Loops for Lap Splice Enhancement Circular Column

Table 3: Maximum Loads on Type I Columns

Column	Maximum Push Load	Max. Pull Load
As Built	33.79 kips (159.19 kN)	35.34 kips (157.19 kN)
Replark30 Jacket	36.26 kips (161.28 kN)	35.43 kips (157.59 kN)
CF Jacket	39.27 kips (174.67 kN)	40.56 kips (180.41 kN)

For Type II columns, the as built column failed in shear at displacement ductility of 1.5. The jacketed columns performed extremely well with displacement ductility of greater than 10. The envelope of the hysteresis obtained is compared in Figure 6. The maximum lateral loads observed are tabulated in Table 4.

Table 4: Maximum Loads on Type II Columns

Column	Max. Push Load	Max. Pull Load
As Built	103.05 kips (458.37 kN)	88.72 kips (394.63 kN)
Replark30 Jacket	114.85 kips (510.85 kN)	109.99 kips (489.24 kN)
CF Jacket	118.75 kips (528.20 kN)	109.52 kips (488.92 kN)

WORK IN PROGRESS

General: Currently a comprehensive study is underway investigating the confinement and strength effects on concrete columns externally reinforced with Fiber Reinforced Polymer (FRP) composite laminates. The experimental phase of this work includes testing of a total of 105 large-scale columns, externally reinforced with both carbon/epoxy and E-glass composite laminates. All specimens will be subjected to axial loading conditions under a quasi-static manner. A 7 million-pound axial testing machine is being used in this program. All the specimens are instrumented with strain gages bonded at both the concrete surfaces, the laminate surface (at two perpendicular directions), and for the reinforced specimens, another group of strain gages were bonded to the steel rebars. In addition, several LVDT's are used to capture the vertical displacements. The feasibility of using laser technology to accurately capture the vertical displacement is being investigated.

The objective of this study is to investigate the size factor, shape of specimen (i.e., circular, square, rectangular, and irregular), confinement ratio, and exiting reinforcement and their effect on the behavior of retrofitted columns using composite jackets. This phase is followed by a detailed analytical study to develop a reliable, general confinement model for columns reinforced externally with composite jackets.

Confined Concrete: The strength and ductility of confined concrete far exceed that of unconfined concrete. Confined concrete is achieved by providing means that would provide lateral pressure exerted on the concrete core to provide additional load capacity. Concrete may be confined by many means some of them are: ties, hoops, spirals, and jackets made of different materials such as steel or composites.

Existing Models: Richard Brantzag & Brown [4] conducted experiments on concrete cylinders laterally confined by fluid pressure. In this study, the fluid pressure, f'_l, was kept constant while the axial stress was increased up to failure. It was found that the confining pressure increased both, the peak concrete stress, f'_{cc}, and the strain at which it was achieved. It was suggested the following equations for the ultimate strength and the corresponding axial strain:

$$f'_{cc} = f'_{co} + 4.1 f'_l \qquad (2)$$

$$\epsilon_{cc} = \epsilon_{co}(1 + 20.5 \frac{f'_l}{f'_{co}}) \qquad (3)$$

where:
f'_{cc} is the strength of confined concrete
ϵ_{cx} is the strain at f'_{cc}
f'_{co} is the strain of unconfined concrete
ϵ_{co} is the strain at f'_{co}
f'_l is the lateral fluid pressure.

Balmer et al. [5] conducted triaxial test on concrete cylinders at lateral confining pressures up to 25,000 psi. For concrete with $f'_c=3{,}573$ psi they suggested the following formula for the strength of the confined concrete:

$$f'_{cc} = f'_{co} + 40.32 f'_l{}^{0.740} \tag{4}$$

All stresses in this equation (4) are in psi.

Mander et al. [6] proposed a model for concrete confined by transverse steel in the form of spirals or hoops. This model is one of the most popular models and is the basis for most current codes. In this model, a stress-strain relationship was developed for confined concrete based on the confining stress and the maximum strength of the confined concrete. The model is based on equation proposed by Popovics 1973 [6] for compressive strength of concrete. Based on this model, the confined concrete compressive strength, f'_{cc}, can be obtained as follows:

$$f'_{cc} = f'_{co}\left(2.254\sqrt{1+\frac{7.94 f'_l}{f'_{co}}} - \frac{2 f'_l}{f'_{co}} - 1.254\right) \tag{5}$$

Mander et al. assumed that failure occurred at the first hoop fracture. The strain at which the first hoop fracture occurs is defined as the ultimate longitudinal strain, ϵ_{cu}. At this strain the column is considered failed. Based on strain energy the following empirical equation for the ultimate longitudinal strain of confined concrete was derived by Priestley et al. [1]:

For circular columns:

$$\epsilon_{cu} = 0.004 + \frac{2.5 \rho_j f_{uj} \epsilon_{uj}}{f'_{cc}} \tag{6-a}$$

For rectangular columns:

$$\epsilon_{cu} = 0.004 + \frac{1.25 \rho_j f_{uj} \epsilon_{uj}}{f'_{cc}} \tag{6-b}$$

where: ρ_j is volume ration of transverse steel to core section

f_{uj} is ultimate jacket strength

ϵ_{uj} is strain at maximum tensile stress in the jacket

f'_{cc} is confined concrete strength

In 1997, Mirmiran and Shahawey [8] have developed a new mathematical model for FRP-encased concrete. This model was obtained only for circular column sections in which the axial response is bilinear with no descending branch. The confined concrete compressive strength, f'_{cc}, is given by:

$$f'_{cc} = f'_{co} + 4.269 f'^{0.587}_l \tag{7}$$

Specimens' Details: The proposed experimental study consists of testing full-scale specimen to develop a new confinement model for concrete confined by composite wraps. A total of 105 specimen are being tested under uni-axial loading conditions in a quasi-static manner. The specimens consist of circular, square, rectangular and irregular shapes as shown in table 5 and 6.

The specimen were confined by FiberBond® Unidirectional Laminates (Manufactured by Edge Structural Composites, LLC) wrapped around the column. Both carbon/epoxy and E-glass/epoxy composites were used. All the specimens were instrumented at mid height with strain gages bonded at both the concrete surface, the laminate surface (at two perpendicular directions), and for the reinforced specimens. Another group of strain gages was bonded to the vertical steel rebars as well as the ties, hoops and spirals as applicable. In addition, several LVDT's were used to capture the vertical displacements. Figure 5 shows the test setup of a typical circular control column specimen.

CONCLUSIONS

Experimental study has demonstrated that advanced carbon fiber composite jackets can significantly improve the ductility performance of old bridge columns with poor lap splice details and insufficient confinements. Such performance can be predicted by the analysis using an equivalent steel jacket. It was also observed from the testing result that the lateral stiffness of the columns are not affected by the composite jackets, while the steel jacket does affect the lateral stiffness thus altering the bridge dynamic characteristics.

The study reported in this paper is a part of a comprehensive test program currently underway at UCI and CSUF. More columns with different cross-sections as well as different types of composite material systems will be tested.

ACKNOWLEDGMENTS

This project was funded by Mitsubishi Corporation through the Society for Advancement of Material and Process Engineering and monitored by the California Department of Transportation. The composite materials for the confinement project were donated by Edge Structural Composites, LLC.

Table 5: Test Program and Properties of Unreinforced Test Specimen

Specimen Cross Section	Specimen Height	No. of Control Specimens	No. of Jacketed Specimen													
			No. of Plies of Carbon Fiber							No. of Plies of E-Glass						
			2	3	4	5	6	8	10	3	4	6	7	8	11	13
Rectangle (10"x15") (25.4 cm x 38.1 cm)	30" (76.2 cm)	3	3	3	-	3	-	3	-	3	3	-	3	-	3	-
Circular (16" ϕ X32") (40.6 cm ϕ x 81.2 cm)	32" (81.2 cm)	3	3	-	3	-	3	-	3	3	-	3	-	3	-	3
Square (15"x15") (38.1 cm x 38.1 cm)	30" (76.2 cm)	3	3	-	3	-	3	-	3	3	-	3	-	3	-	3
Irregular (Hexagonal)	30" (76.2 cm)	1	-	-	3	-	-	-	-	-	-	-	-	-	-	-
Irregular (Octagonal)	30" (76.2 cm)	1	-	-	3	-	-	-	-	-	-	-	-	-	-	-

Table 6: Specimen with Reinforcing Steel

Specimen's Cross Section	Vertical Steel Reinforcement		NO. OF REINFORCED SPECIMEN			
			No. of Control Specimens	No. Of Jacketed Specimen using 4-Ply Of Carbon Fiber		
	Bar selection	ρ_s (%)		#2 Ties @ 3" o.c.	#2 Hoops @ 3" o.c.	#2 spiral w/ 3" pitch
Rectangular (10"x15")	8#5	1.6	1	2	-	-
Circular (16" ϕ X32")	10#5	1.5	1	-	2	-
	10#5	1.5	1	-	-	2
Square (15"x15")	12#5	1.6	1	2	-	-

Figure 5: Test Setup of a Typical Circular Control Column Specimen in the 7 Million-Pound Testing Machine.

REFERENCES

1. Priestley, M.J.N., Seible, F. and Calvi, G.M., *Seismic Design and Retrofit of Bridges*, John Wiley and Sons, Inc., 1996.
2. Xiao, Y., Martin, G.R., Yin, Z. and Ma, R., *Bridge Column Retrofit using Snap-Tite Composite Jacketing for Improved Seismic Performance*, University of Southern California, Structural Engineering Research Report No USC-SERR95/02, June 1995.
3. Priestley, M.J.N., Seible, F. and Fyfe, E., *Column Seismic Retrofit using Fiberglass/Epoxy Jackets*, Proceedings ACMBS-1 Conference, Quebec, Canada, October 1992, pp 287-297.
4. Richart, F.E., Brantzaeg, A. and Brown, R.L. (1928). *"A Study of the Failure of Concrete under Combined Compressive Stresses"*. Bulletin No. 185, Engineering Experiment Station, University of Illinois, Urbana, IL, 104p.
5. Balmer, G.G., Jones, V., and McHenry, D. (1949). *"Shearing Strength of Concrete under High Triaxial Stress-Computation of Mohr's Envelope as a Curve"*. Struct. Res. Lab. Report No. SP 23, Research and Geology Division, US Bureau of Reclamation, Oct., 27p.
6. S. Popovics, *"A Numerical Approach to The Complete Stress-Strain Curves For Concrete,"* Cement and Concrete Research, Vol. 3, No. 5, pp. 583-599, 1973
7. T. Paulay and M. Priestley, Seismic Design of Reinforced Concrete and Masonry Buildings. John Wiley & Sons, Inc., 1992.
8. Mirmiran, A., and Shahawy, M. *"Analytical and Experimental Investigation of Reinforced Concrete Columns Encased in Fiberglass Tabular Jacket and Use of Fiber Jacket for Pile Splicing."* University of Central Florida, Final Report, Feb. 1997.
9. Mander, J. B., M. N. Priestley, and R. Park, *"Theoretical Stress-Strain Model for Confined Concrete."* Journal of the Structural Division, ASCE, Vol. 114, NO. 8, August 1988, pp. 1804-1826.
10. Mander, J. B., M. N. Priestley, and R. Park, *"Observed Sets-Strain Behavior of Confined Concrete,"* Journal of the Structural Division, ASCE, Vol. 114, NO. 8, August 1988, pp. 1827-1849.

The Retrofit Design of Concrete Columns and Slabs with Externally Applied Fiber-Reinforced Polymer (FRP) Composite Materials

R. M. ELHASSAN

ABSTRACT

Externally applied Fiber-Reinforced Polymer (FRP) composite materials can significantly enhance the structural performance of existing structural elements by enhancing their strength and deformation capacities under various static and dynamic loading conditions, including seismic and gravity effects. These materials have been used successfully with existing concrete and masonry elements. However, due to the lack of design criteria and code requirements for the use of these materials, structural engineers have either relied on the design services of FRP material suppliers, which is the most common, or have developed project-specific design criteria and procedures based on their own research and experience. This paper illustrates the design and analysis procedures used in two projects to strengthen existing columns and slabs with externally applied FRP materials to resist the imposed gravity and earthquake loads.

INTRODUCTION

The use of Fiber-Reinforced Polymer (FRP) composite materials in the repair and retrofit projects has been steadily increasing for the past several years as external reinforcement to existing concrete and masonry structural members. However, their use has been hindered by the lack of design criteria and codes that provide prescriptive requirements for their design.

FRP materials can be distinctly classified as Glass-fiber (GFRP) or Carbon-fiber (CFRP) composite materials based on the type of their primary fiber. However, the retrofit design of elements externally reinforced with these materials can be accomplished in general by considering FRP layers as additional tension or compression reinforcement with the following distinct material characteristics:

President and Principal Engineer, BFL Owen & Associates, 5 Goodyear Street, Irvine, California 92618, USA; and, Chairman, Fiber-Reinforced Polymer Composite Committee of the Structural Engineers Association of California

1. Tensile stress-strain behavior of FRP materials is highly linear up to failure point.

2. Compared to steel, FRP materials are brittle, less stiff (smaller Modulus of Elasticity), and have much smaller strain capacity.

3. Unconfined, FRP systems do not sustain compression, and may buckle and debond under low compression load.

The assessment of the strength and deformation capacities of the existing elements should be established first and their deficiencies defined. FRP materials are then designed to supplement existing capacities to resist imposed loading demands, taking into consideration the existing state of stresses, strains and deformations. The strain and stress "design" values of FRP materials are related to the nominal stress and strain values by "capacity reduction factors" that are specifically established for the type of element, load, and FRP fiber used, and for the design limit-state. Also, the design stress or strain of the FRP materials is limited by the acceptable limits of the substrate.

Table 1 shows typical properties of dry fibers, and Table 2 shows typical properties of FRP composite systems, which consist of the dry fibers saturated with resin. Dry fiber properties vary greatly depending on the fiber type, fiber grade, and the aerial weight/density of fiber. FRP system properties can vary depending on the manufacturer and on other factors such as fiber-to-resin ratio, fiber properties, resin properties, aerial weight/density of fiber, and type of weave/stitch.

TABLE 1. TYPICAL PROPERTIES OF GLASS AND CARBON DRY FIBERS

PRIMARY FIBER	GLASS	CARBON
Tensile Strength	270-700 ksi	350-700 ksi
Ultimate Strain	>0.04	>0.005
Elastic Modulus	10,000 ksi	30,000 ksi

TABLE 2. TYPICAL PROPERTIES OF GLASS AND CARBON COMPOSITE SYSTEMS

PRIMARY FIBER	GLASS	CARBON
Tensile Strength (min, ASTM D3039)	60-80 ksi	100-150 ksi
Ultimate Strain	0.015-0.022	0.012-0.017
Elastic Modulus (min. ASTM D3039)	3,000-4,000 ksi	8,000-12,000 ksi
STRENGTH AT 90°	5.0 ksi	5.0 ksi
Coefficient of Thermal Expansion	4.3E-6/°F	0.24E-6/°F

PERFORMANCE-BASED SEISMIC RETROFIT OF PARKING STRUCTURES USING GFRP

This project consists of two parking structures that are situated to the sides of the terminal building of a major Southern California airport. Each parking structure consists of two structural portions that are separated by seismic joint, thus, the project consists of four individual structural portions, with a combined total floor area of 1,100,000 sq.ft over four levels. The structures were designed and built in 1988-1991, using the 1985 Edition of the Uniform Building Code (UBC).

The structural/seismic evaluation performed concluded that critical structural deficiencies existed in the gravity columns, which cannot withstand the lateral inter-story movement of the structure during a strong seismic shacking. Thus, the scope of this project consisted of the design of a structural remediation scheme to correct the seismic deficiencies and improve the seismic response of the structures to future earthquakes. The structures are considered essential facilities.

Design Criteria and Computer Modeling and Analysis

The "Guidelines for the Seismic Rehabilitation of Buildings" (FEMA 273), published in 1997 by the Federal Emergency Management Agency, was used as the basis for the project-specific design criteria prepared for this project. The design criteria required a dual-level performance-based design with specific seismic performance objective for each earthquake level. Two level of earthquake ground motion were used, BSE-1 and BSE-2, as follows:

- BSE-1 (Basic Safety Earthquake-Level 1): This is an earthquake ground motion that has a 10% probability of being exceeded in 50 years, corresponding to a 475-year return period.

- BSE-2 (Basic Safety Earthquake-Level 2): This is an earthquake that has a 2% probability of being exceeded in 50 years, corresponding to a 2500-year return period.

The earthquake performance objectives for the structures were "Immediate Occupancy Performance" for BSE-1 and "Life Safety Performance" for BSE-2.

Detailed computer models were constructed for the parking structures using the computer program SAP2000, and response spectra and time-history analyses were performed using BSE-1 and BSE-2 ground motions. Figure 1 shows the SAP2000 model for one of the parking portions, and Figure 2 shows the 5% damped acceleration response spectra for BSE-1 and BSE-2. The computer models incorporated the primary lateral resisting structural elements (moment frame beams and columns) and all the secondary structural elements (ramps and gravity beams and columns). The force and displacement response were obtained at BSE-1 and BSE-2 earthquakes for all elements using the procedure and the amplification factors of FEMA 273.

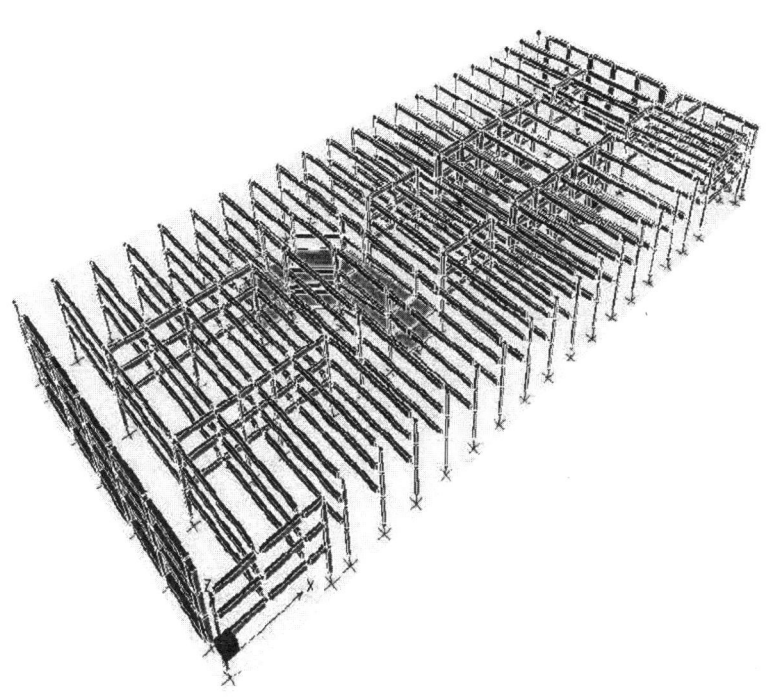

FIGURE 1. SAP2000 COMPUTER MODEL FOR ONE OF THE FOUR PARKING STRUCTURE AREAS

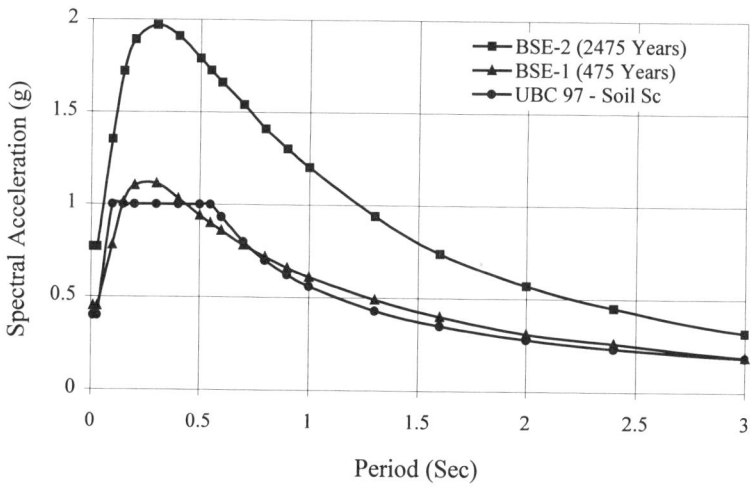

FIGURE 2. FIVE-PERCENT DAMPED ACCELERATION RESPONSE SPECTRA FOR BSE-1, BSE-2, AND UBC-97 FOR SOIL TYPE S_C

Strengthening Scheme and the Retrofit of Columns Using FRP

Several retrofit schemes were examined that included variations of adding shear walls and jacketing of columns. The selected seismic retrofit scheme consisted of FRP column-jacketing and also included widening of the existing seismic separating gaps at the roof level and at the access bridges, as well as special treatment of the ramp columns and stair wall enclosures. This scheme did not include the addition of any shear walls, and provided the desired seismic performance with the shortest construction schedule, the least interruption to parking operation during construction, and a minimal impact on the architectural configuration of the retrofitted structures.

Site structural survey and destructive and non-destructive testing were conducted on the existing structures to verify their as-built material properties and as-built sizes. Non-linear moment-curvature analyses were performed on all the columns using the mean values of the tested properties, and their moment, shear, and lateral deformation capacities were determined. The force, moment and lateral deformation demands on various columns were obtained from the computer analyses and were compared to their existing strengths and deformation capacities according to the requirements of FEMA 273 criteria. Where the existing shear or moment capacities did no satisfy these requirements for a column, then the column was retrofitted using FRP jacket.

The project-specific FRP design criteria consisted of the following:

- For Immediate Occupancy Performance Level to BSE-1 and Life-safety Performance Level to BSE-2, the column shear and moment strengths were checked against the shear and moment demands using FEMA 273 criteria for the stated performance objective.

- When the FEMA 273 prescriptive criteria are not met then a limit state analysis was used to design the FRP jackets to provide the desired performance objective. For a column, the shear force demand corresponding to the development of the plastic moment strength at the column top and bottom ends was calculated and set as the shear force demand on the column. And, the ultimate lateral displacement demand on the column was determined from the computer analysis. The FRP jacket was then designed to insure that the column shear strength is higher than the shear force demand, and the column lateral deformation capacity was larger than the ultimate deformation demand.

The column FRP retrofit was carried out using glass-fiber polymer composite materials, or GFRP, that formed passive continuous jacket around the column, where the main fibers were oriented in the horizontal (transverse) direction. The GFRP jacket resists the shear force in the column by restraining the formation of flexural-shear cracks, and provides confinement pressure that increase the ultimate concrete compressive strain and thus enhance the lateral deformation capacity of the column. Uniform jacket thickness was used through the column clear height for shear strengthening, and thicker jackets were used at the column top and bottom regions to confine the potential plastic hinge regions.

Shear Strengthening of Columns:

Limit state analysis was used to eliminate shear failure and to allow the columns to undergo post-yield ductile flexural yielding. The column shear force corresponding to the development of plastic hinge at the top and bottom of the column clear height was obtained using an overstrength factor of 1.25 with the tested mean properties. The column shear strength was calculated using the recommendations of Reference 1, which considers the inclination of the flexural-shear cracks to be 35°, and predicts the shear strength of column with varying degree of ductility demands. Thus, the shear strength "design" equations of an existing column section based on Reference 1 is:

$$Vc + Vs + Vp = 0.8 \ k \ \sqrt{f'c} \ bh + 1.43 \ Av \ fyv \ d'/s + 0.85 \ P \ (h-cp)/L_{clr} \quad (1)$$

Where:

$Vc =$ Concrete contribution to shear strength
$Vs =$ Steel reinforcement (truss mechanism) contribution to shear strength
$Vp =$ Axial load contribution to shear strength
$k =$ Concrete shear strength contribution factor that ranges from 3.0 for column lateral displacement ductility (μ_Δ)<2 to 0.6 for μ_Δ>8 at the plastic hinge region, and a value of 3.0 at the column central region
$f'c =$ Concrete compressive strength
$b =$ Width of section
$h =$ Depth of section
$Av =$ Area of tie/hoop rebars
$fyv =$ Yield stress of tie/hoop rebars
$d' =$ Effective depth of section
$s =$ Spacing of ties/hoops
$P =$ Column axial load
$L_{clr} =$ Clear height of column
$cp =$ Depth of neutral axis at nominal moment strength

The shear strength provided by the jacket is then calculated based on a flexural-shear crack inclination angle of 35°, as follows:

$$Vj = 2.86 \ (n_p \ tj) \ \varepsilon_v \ (C_{EM} \ Ej) \ h \quad (2)$$

Where:

$n_p =$ Number of FRP layers for shear
$tj =$ Thickness of one layer of FRP jacket
$\varepsilon v =$ Design tension strain of FRP jacket for shear strength, equal to 0.004

C_{EM} = Environmental Reduction Factor to account for the long-term degradation of the Modulus of Elasticity of the jacket due to environmental exposure, equal to 0.90
E_j = Nominal Modulus of Elasticity of FRP jacket

The total shear strength of the column is then calculated for the mid-region of the column height and for the plastic hinge regions as follows:

$$V_u = \phi \, (V_c + V_s + V_p + \alpha_j V_j) \quad (3)$$

Where:

V_u = Shear strength
ϕ = Capacity Reduction Factor for shear strength, per ACI-318
α_j = Capacity Reduction Factor for the FRP contribution, taken as 0.85

Confinement of Plastic Hinge Regions

The lateral displacement capacity of the existing column was compared to the displacement demand from the dynamic analysis and confinement of the plastic hinge regions at the column's top and bottom was carried out as needed to increase the column's lateral displacement capacity. Strengthening for shear, as required in the previous paragraph, eliminated shear failure of the column.

The yield curvature and displacement (ϕy and Δy) of the existing column, which correspond to the first yield of its flexural reinforcement, and the unconfined curvature and displacement capacities (ϕuu and Δuu), which correspond to a maximum concrete compressive strain of 0.004, were calculated using the results of the moment-curvature analysis. The unconfined displacement capacity of the column is calculated assuming that plastic hinges are formed at the column top and bottom ends, and based on a length of the plastic hinge region using Reference 1. The unconfined displacement ductility (μ_Δ) of the column was calculated as $\Delta uu/\Delta y$.

Where the unconfined lateral displacement capacity of the column was less than the lateral displacement demand, then confinement of the plastic hinge regions was carried out by wrapping them with GFRP, with the main fibers oriented transversely to the column longitudinal axis. The GFRP confinement ratio, ρ_j, was calculated as:

$$\rho_j = 2 \, n_c \, t_j \, (b+h) / bh \quad (4)$$

Where n_c is the total number of GFRP layers in the plastic hinge region. The confined concrete compressive strength, f'cc, for the rectangular concrete column sections was assumed to be 1.5f'c. Thus, the maximum confined concrete compressive strain is calculated as:

$$\varepsilon u = 0.004 + \alpha_j (1.25 \, \rho_j \, C_E \, f_{uj} \, \varepsilon_{uj} / f'cc) \quad (5)$$

Where f_{uj} and ε_{uj} are the nominal ultimate tensile stress and strain of the FRP jacket, α_j is the Capacity Reduction Factor for the FRP contribution (taken as 0.85), and C_E is an ultimate Strength Reduction Factor to account for the long-term degradation of ultimate strength of the jacket due to environmental exposure, taken as 0.75.

The confined lateral displacement capacity, Δu, is then calculated and compared to the lateral displacement demand, and the curvature capacity, ϕu, and the curvature and displacement ductility capacities, $\mu\phi = \phi u / \phi y$, and $\mu\Delta = \Delta u / \Delta y$ are calculated. Figure 3 compares the moment-curvature diagram for an unconfined column with that of the same column using 4 confinement layers of GFRP at the column plastic hinge regions.

A total of more than 700 columns were retrofitted as part this project using an average of three to four layers of GFRP.

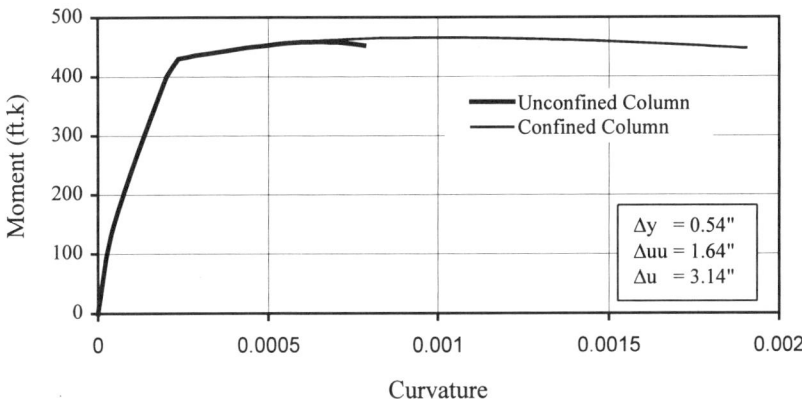

FIGURE 3. MOMENT-CURVATURE PLOTS FOR 18"X24" COLUMN, UNCONFINED AND CONFINED WITH 3 LAYERS OF GFRP

RETROFIT OF A CONCRETE SLAB USING CFRP

This project consists of the upgrade of the vertical load-carrying capacity of an existing concrete flat-plate of an office building located in San Jose, California. The objective of the retrofit was to increase the live-load carrying capacity of the existing concrete slabs from a design load of 50 psf to 100-150 psf. The live load requirements were changed due to an ownership transfer that resulted in a change of use/occupancy of the building. The column spacing of a typical bay is 32ft.x27ft., and the existing slab thickness ranges from 11 to 14 inches.

The new ultimate moment demands was calculated and compared to the existing moment strengths using two-way flat plate design program. Where moment deficiencies were noted, the new service, Ms, and ultimate, Mu, moment demands were used along with the existing steel reinforcement and material properties to design the required CFRP reinforcement. In addition, the in-place moment, Mi, which is the moment calculated based on the dead-plus-live loads expected to exist at the time of the CFRP installation, was also calculated. Only positive moment deficiencies existed in this project.

Design of FRP Reinforcement

The "nominal" FRP tensile strength (fuj) and Modulus of Elasticity (Ej) values reported by the manufacturer were reduced for the retrofit "design" by the Environmental Reduction Factors, C_E and C_{EM}, to account for the long-term degradation of ultimate strength and Modulus of Elasticity of the FRP due to environmental exposure. A design value for the rupture strain, εuj, is calculated based on the design values of the fuj and Ej. The "design" values replace the "nominal" values wherever required in the design below, where C_E and C_{EM} were taken as 0.90 and 0.95 respectively.

The design procedure for the CFRP retrofit of these concrete slabs included computing five states of stress and strain or limit states, as follows:

- *Initial State of Stress and Strain:* This is based on the in-place moment at the time of installation of the CFRP. The locked-in initial concrete and steel stresses and strains were calculated. There are no design requirements at this state.

- *Service-load Limit State*: This is based on the retrofitted section with the full dead and live loads. Concrete, steel and FRP stresses (fc, fs, fj) for the applied Ms moment were checked versus the allowable stresses for the concrete, steel, and FRP as follows:

$$fc < Rc\ f'c \tag{6}$$

$$fs < Rs\ f'y \tag{7}$$

$$fj < Rf\ fuj \tag{8}$$

Where Rc=0.45, Rs=0.80, and Rf=0.33 are the service load allowable stress factors for the concrete, steel, and FRP materials (for FRP, Rf is based on creep rupture limit state with a factor of safety)

- *Steel-yield Limit State:* This limit state establishes the moment at the initial yield of rebars, My, of the retrofitted section. The moment Factor of Safety against yielding of steel reinforcement is calculated. No minimum requirement has been set for this project for that Factor of Safety.

- *Ultimate-load Limit State:* Two limit states and their corresponding ultimate moment strengths are established: the "concrete-crushing" limit state establishes the ultimate moment strength of the retrofitted section, Mnc, based on a concrete strain of εuc=0.003, and the "FRP-rupture" limit state establishes the ultimate moment strength of the retrofitted section, Mnj, based on a FRP "design" stress of fuj. In the ultimate limit state moment calculations, the FRP contribution to the total moment strength was reduced by the FRP Capacity Reduction Factor αj=0.85. For the ultimate limit state, the moment strength, Mn, shall be as follows:

$$Mn = Min\ (Mnc, Mnj) \qquad (9)$$

$$\phi\ Mn > Mu \qquad (10)$$

Where $\phi = 0.90$ is the beam Capacity Reduction Factor per ACI-318.

For the FRP reinforced sections, the moment-curvature diagrams, the yield and ultimate curvatures (ϕy, ϕu), and the curvature ductilities ($\mu\phi = \phi u\ /\ \phi y$) were obtained and compared to those of the sections without FRP reinforcement.

A parametric study was performed to address the reduction of the curvature and the curvature ductility capacities of the FRP reinforced sections. It was concluded that the reduction in the ultimate curvature capacity is related to the amount of existing steel reinforcement (reinforcement ratio, ρs) and the amount of FRP reinforcement added. Figure 4 shows the moment-curvature diagrams of a reinforced concrete section (18"x24") with various steel reinforcement ratios without FRP (shown as the bottom curve in every case, termed "existing") and with varying amount of FRP reinforcement (the curves termed FRP1, FRP2, FRP3 and FRP4, where the amount of FRP reinforcement in FRP4 is double that in FRP3, and in FRP3 is double that in FRP2, and in FRP2 is double that in FRP1). It was concluded that:

- For lightly steel-reinforced sections, large ultimate strength enhancement is realized with the use of FRP reinforcement, but the ultimate curvature capacity is significantly reduced for the sections with FRP, and the reduction increases proportionally with the amount of FRP added.

- For heavily steel-reinforced sections, only slight ultimate strength enhancement is realized with the use of FRP reinforcement, but the ultimate curvature capacity is slightly reduced for sections with FRP, and the reduction increases proportionally with the amount of FRP added.

- The ultimate curvature capacity of the FRP reinforced section decreases as the governing ultimate limit state changes from FRP-rupture to concrete-crushing limit state.

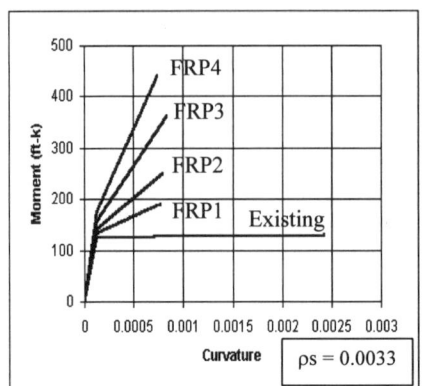
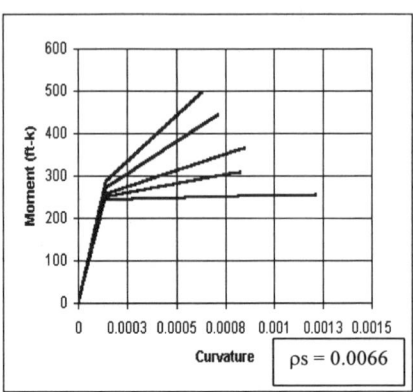
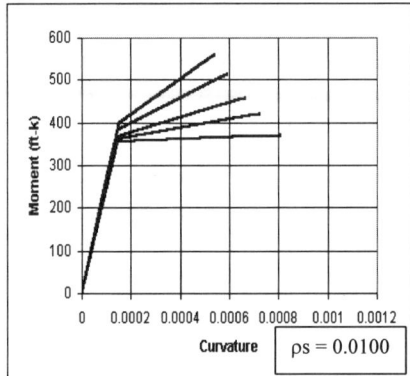
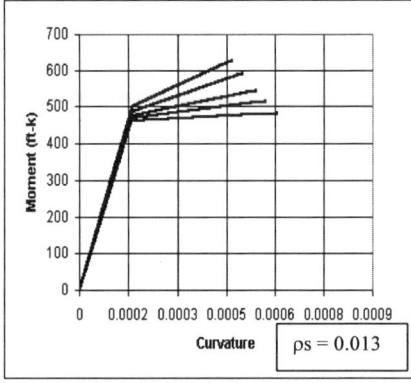
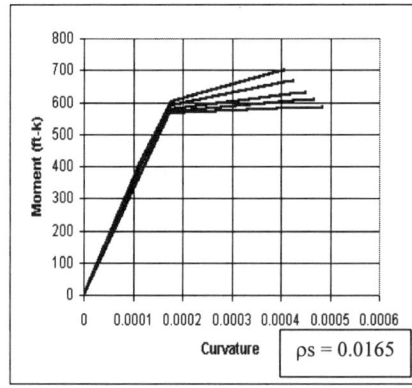
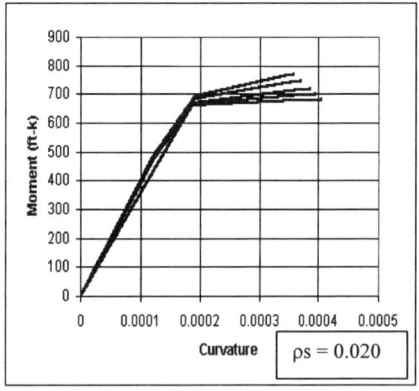

FIGURE 4. MOMENT-CURVATURE DIAGRAMS OF A RECTANGULAR SECTION WITH VARIOUS STEEL REINFORCEMENT RATIOS, WITHOUT FRP AND WITH VARYING AMOUNT OF FRP REINFORCEMENT

A total of 28 slabs required retrofit in this project, which ranged from providing 1-ply, 10-inch wide strips spaced 30" apart, to 2-ply, 10-inch wide strips spaced 1" apart (clear spacing).

CONCLUSIONS AND RECOMMENDATIONS

Fiber-reinforced Polymer composite materials (FRP) can significantly enhance the structural performance of existing structural elements by enhancing their strength, stiffness, and ductility. There are many uses of these systems in the retrofit of existing structural members as external reinforcement to resist various static and dynamic loading conditions, including seismic and gravity effects.

While the use of FRP materials is finding acceptance in the structural engineering community it is still in its infancy, and there is still significant experimental and analytical work needed to better quantify their design properties and procedure. There is also still a great need for design criteria and codes that provide prescriptive design requirements for these systems.

ACKNOWLEDGMENT

The author wishes to acknowledge the contribution of Mr. T.J. (Touraj) Eimani, Ph.D., SE, of BFL Owen & Associates in the project engineering, computer modeling and analysis of the seismic retrofit of the parking structures presented in this paper.

REFERENCES

1. Priestley, M. J. N., Seible, F., and G. M. Calvi. 1996. *Seismic Design and Retrofit of Bridges*. New York, USA: John Wiley and Sons, Inc.

2. American Concrete Institute. 1996. "State-of-the-Art Report on Fiber Reinforced Plastic (FRP) Reinforcement for Concrete Structures," ACI Committee 440 Report.

3. International Conference of Building Officials. 1997. "Acceptance Criteria for Concrete and Reinforced and Unreinforced Masonry Strengthening Using Fiber-Reinforced, Composite Systems," AC125 Report, ICBO Evaluation Service, Inc., Whittier, California.

Prediction of Cyclic Performance of Composite-Jacketed Squat Reinforced Concrete Bridge Columns

M. A. HAROUN[1] and H. M. ELSANADEDY[2]

ABSTRACT

Ten half-scale squat circular and rectangular reinforced concrete columns were tested in shear to evaluate the effect of composite jacketing as a retrofit measure to older bridge columns. Two columns were tested in the as-built condition, while eight columns were tested after being retrofitted with different composite-jacket systems. The as-built columns failed in brittle shear mode, whereas the retrofitted columns underwent flexural failure modes with significant improvement in the column ductility. The emphasis of this research is on macroscopic modeling of all tested columns. Such a model was employed to predict the behavior of the as-built columns by evaluating different shear strength models using test results of 57 shear-deficient columns available in the literature. The model was also calibrated to evaluate the performance of composite-jacketed columns through a parametric study on two displacement models and six different concrete confinement models.

INTRODUCTION

One of the major problems associated with the seismic performance of reinforced concrete bridges is the brittle shear failure of squat bridge columns. Such short and, hence, relatively stiff members tend to attract a greater portion of the seismic input to the bridge during an earthquake and require the generation of large seismic shear forces to develop the moment capacity of columns. Estimation of flexural strength based on elastic methods, along with much less conservative shear strength provisions during the 1950s and 1960s, frequently resulted in actual shear strength of as-built bridge columns being significantly less than the flexural capacity. Generally the transverse reinforcing steel was inadequately anchored in the cover concrete, which can be expected to spall off under cyclic loading, and therefore the problem was compounded. Hence, shear failure is likely in such columns, accompanied not only by rapid strength, stiffness, and physical

---

[1]Professor and [2]Graduate Research Assistant, Department of Civil and Environmental Engineering, University of California, Irvine, CA 92697

degradation but also by energy dissipation characteristics. This has been evidenced by the brittle shear failure of bridge columns in recent California earthquakes.

A full-height steel jacketing method for retrofitting squat bridge columns has been proven experimentally to be effective in alleviating shear failure problems and enhancing their seismic performance. Recently, advanced composite materials have shown great potential to becoming a viable alternative to steel jackets. The light weight, high strength or stiffness-to-weight ratios, corrosion resistance, engineered properties, and more importantly the ease of installation make such materials most suitable for retrofitting bridge columns.

In order to examine and qualify the seismic performance of squat bridge columns retrofitted with composite-material jackets, an experimental study was conducted at the University of California, Irvine, on both circular and rectangular columns [7, 8]. These columns have been tested for shear enhancement in a fixed-fixed condition.

DETAILS OF TEST SPECIMENS

Ten half-scale columns have been tested in shear; one circular as-built column (CS-A1), two circular columns retrofitted by two different jackets (CS-R1 and CS-R2), one rectangular as-built column (RS-A1), and six rectangular columns retrofitted by six different types of jackets (RS-R1 to RS-R6). It should be noted that the letters 'C' and 'R' denote circular and rectangular columns, respectively, the letter 'S' denote shear testing, and the letters 'A' and 'R' denote as-built and retrofitted columns, respectively. All columns had an aspect ratio of 2 with a clear height of 96 inch (2.44 m), and were reinforced longitudinally by 20#6 (d_b = 1.91 cm), uniformly distributed around the column section with a concrete cover of 1 inch (2.54 cm). Circular columns had a diameter of 24 inch (60.96 cm) and were transversely reinforced by #2 hoops (d_b = 0.64 cm) spaced at 5 inch (12.7 cm). Rectangular columns were built with an 18 inch x 24 inch (45.72 cm x 60.96 cm) rectangular cross-section, and had transverse steel of #2 ties (d_b = 0.64 cm) spaced at 5 inch (12.7 cm). All columns were reinforced with grade 40 steel (nominal strength = 40 ksi = 276 MPa) and had a nominal concrete strength of 5 ksi (34.5 MPa). Details of all tested specimens are shown in Table I. It should be noted that the composite jacket properties in Table I are average values based on the test results of control samples.

TEST SET-UP AND PROCEDURE

Test Set-up

The test set-up was designed to subject the columns to a constant axial compressive load and cyclic horizontal loads. As shown in Fig. 1, each of the shear columns was subjected to a lateral load applied by a specially-designed test rig using a pantagraph arrangement to restrain the rotation at the top of the column, thereby simulating the fixed-fixed condition

Test Procedure

An axial load was applied to each column specimen by post-tensioning two steel rods with a hydraulic jack at the top of the column. According to Caltrans guidelines [5], peak force controls the initial loading cycles for each test until the column developed the lateral load corresponding to the first yield of longitudinal steel, V_y. Then, the test was stopped and the yield displacement was determined from

$$\Delta_y = \frac{V_i}{V_y}\Delta_1 \qquad (1)$$

where Δ_1 is the average of the measured peak displacements corresponding to the first-yield lateral load, V_y, in the push and pull directions. The ideal flexural lateral load capacity, V_i, is computed based on the extreme concrete compressive strain of 0.004 (0.005 for the jacketed columns) and on measured material properties. After the column developed the first yield capacity, loading cycles were controlled by the peak displacement.

TABLE I. PROPERTIES OF SHEAR ENHANCEMENT COLUMN SPECIMENS

Test Unit	Concrete Strength f'_c (ksi)	Longitudinal Steel f_y (ksi)	Composite Jacket Properties			
			Type	Thickness t_j (inch)	Tensile Strength f_{ju} (ksi)	Tensile Modulus E_j (Msi)
CS-A1	5.34	43.41	As-built circular column			
CS-R1	5.92	43.41	Carbon fiber	0.0264	605	33.60
CS-R2	5.69	43.41	Carbon fiber	0.0264	643	33.40
RS-A1	5.38	43.41	As-built rectangular column			
RS-R1	5.53	43.41	Carbon fiber	0.0396	636	32.80
RS-R2	5.71	43.41	Carbon fiber	0.0399	643	33.40
RS-R3	6.38	43.41	Carbon fiber	0.0396	605	33.60
RS-R4	6.38	43.41	E-glass	0.30	108	5.30
RS-R5	6.38	43.41	Carbon fiber	0.205	136	9.15
RS-R6	6.18	43.41	E-glass	0.30	93	5.29

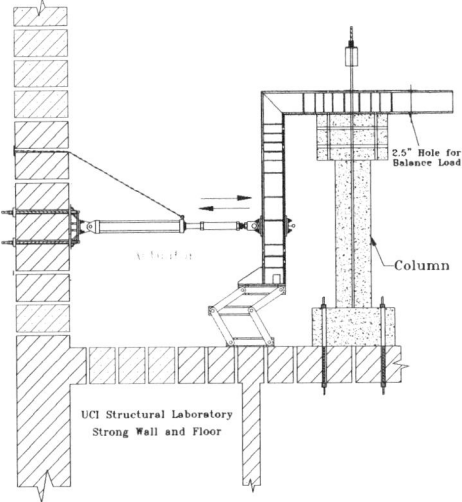

Figure 1. Test Set-up for Shear Enhancement Columns.

TEST RESULTS

The test results in terms of load-displacement envelopes for all 10 columns are shown in Fig. 2 and Fig. 3. Brittle shear failure was observed in the two as-built columns. For specimen CS-A1, shear failure occurred at ductility 1.4 with a shear failure plane at an angle of 30° to the column longitudinal axis. Specimen RS-A1 failed at ductility 0.8 with shear cracking angle of 30° to the column axis. All retrofitted circular columns showed very stable loops and significantly increased ductility without strength degradation. For retrofitted rectangular columns, tests have shown significant improvement in the column ductility.

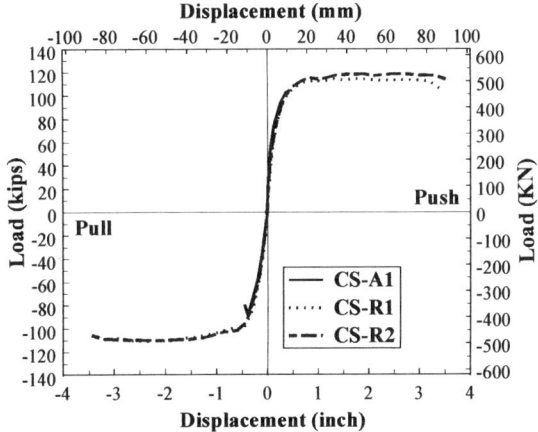

Figure 2. Load-Displacement Envelops for Circular Shear Columns.

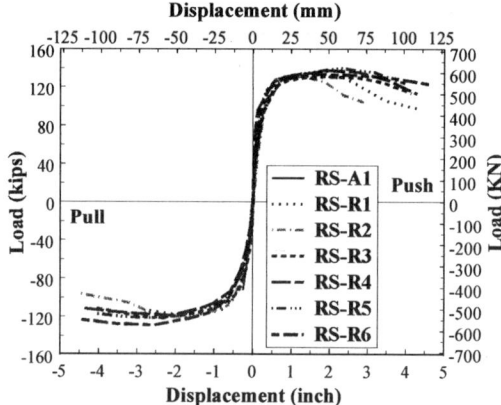

Figure 3. Load-Displacement Envelops for Rectangular Shear Columns.

ANALYTICAL MODELING

An object-oriented computer program, based on moment-curvature analysis, was used to analyze both circular and rectangular shear columns. In defining the significant variables that may affect the performance of the column, the constitutive properties of the materials are of key concern. For as-built columns, concrete was considered unconfined and the stress-strain model for unconfined concrete was used. However, for retrofitted columns, confined concrete models were utilized. Six confinement models were employed in the program. The first is the model developed by Mander et al. [13] which has been successfully applied to both circular and rectangular steel-jacketed columns. The second model was developed by Samman et al. [15] specifically for circular composite-jacketed columns. The third model, recently developed by Hosotani et al. [10], is applicable for both circular and rectangular composite-jacketed columns. The fourth model was suggested by Hoppel et al. [9] for circular columns only. Toutanji [17] and Spoelstra et al. [16], respectively, developed two other models limited to circular composite-jacketed columns. In addition to defining the stress-strain characteristics of concrete, a model of the stress-strain properties of steel reinforcement was employed. This model was divided into three major zones: a linear portion up to the yield, a yield plateau region, and a parabolic strain-hardening region.

The column was analyzed by using laminar analysis procedure for its cross-section in order to obtain the moment-curvature curve. The load-displacement response was obtained by modeling the column as two vertical cantilevers. For each cantilever, two models were utilized for displacement calculation. The first was developed by Kowalsky et al. [12], while the second was suggested by Wehbe et al. [18].

Included in the modeling were the calculations of shear strength envelope for as-built columns. Two sets of shear strength provisions were utilized. For the first set, the concrete shear strength contribution degrades as the column displacement ductility increases. Such approach is used in UCSD shear strength model [14],

Caltrans model [6], UCB shear strength model [3], and Architectural Institute of Japan (AIJ) seismic design guidelines [2]. The second set includes shear strength provisions at which the concrete contribution to shear strength does not depend on the column displacement ductility such as ACI 318-95 provisions [1], ASCE/ACI 426 proposals [11], and ATC-32 provisions [4].

MODEL CALIBRATION

As-Built Columns

In order to predict the performance of as-built columns, the model was used to obtain the flexural strength envelope and shear strength envelope based on different shear strength provisions. The intersection point of the two envelopes determines the maximum shear force and the corresponding ductility. For specimen CS-A1, a comparison of the experimental to the predicted behavior is shown in Fig. 4. It is noticed that the best shear strength model to match the experimental data is provided by Caltrans shear model. As indicated in Fig. 5, the best shear models to fit the experimental data for specimen RS-A1 are Caltrans shear model and AIJ design guidelines.

In order to calibrate the model for the prediction of performance of shear-deficient columns, the current shear strength provisions should be evaluated. For this purpose, a large set of experimental data was collected from different tests, available in the literature, on both circular and rectangular shear-deficient columns. Incorporating the two as-built columns CS-A1 and RS-A1, the total test population has 45 circular columns and 12 rectangular columns. Altogether, there were 57 columns identified as having failures strongly influenced by shear. Theoretical shear strength values of all columns were calculated based on the different shear strength models. The ratio of the experimental shear strength, V_{exp} to theoretical shear strength, V_{th} of each column was computed for all shear models. Statistical summaries of the (V_{exp}/V_{th}) ratios for all shear models are presented in Table II. It is evident that the ATC-32 shear strength model is the most conservative model since it has the highest average (V_{exp}/V_{th}) ratio. From the table, it is also evident that the best fit to the data in terms of both average (V_{exp}/V_{th}) ratio and standard deviation is provided by the UCSD shear model. Based on the statistical lower bounds, applying 0.85 factor to the shear strength provided by the UCSD model provides a reasonable design value for shear strength calculations.

TABLE II. STATISTICAL ANALYSIS FOR SHEAR-DEFICIENT COLUMNS

Model	m	σ	m - σ	m − 2σ
UCSD	1.065	0.106	0.959	0.853
Caltrans	1.271	0.252	1.019	0.767
UCB	1.089	0.146	0.943	0.797
AIJ	1.143	0.225	0.918	0.693
ACI	1.375	0.343	1.032	0.688
ASCE-ACI	1.306	0.302	1.004	0.702
ATC-32	2.052	0.629	1.423	0.795

Note: m: average value; σ: Standard deviation; m - σ and m − 2σ: Statistical lower bounds

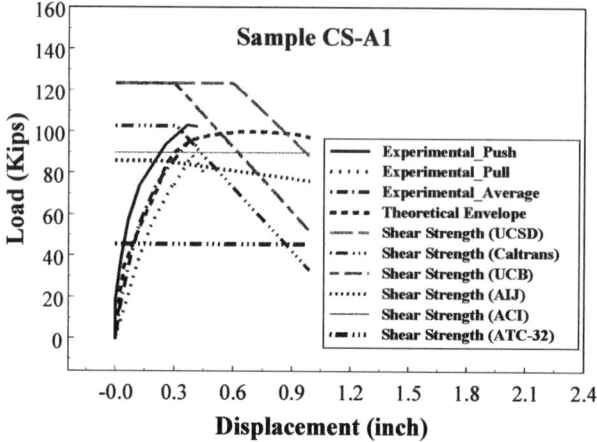

Figure 4. Experimental and Theoretical Comparison of Shear Response of Specimen CS-A1.

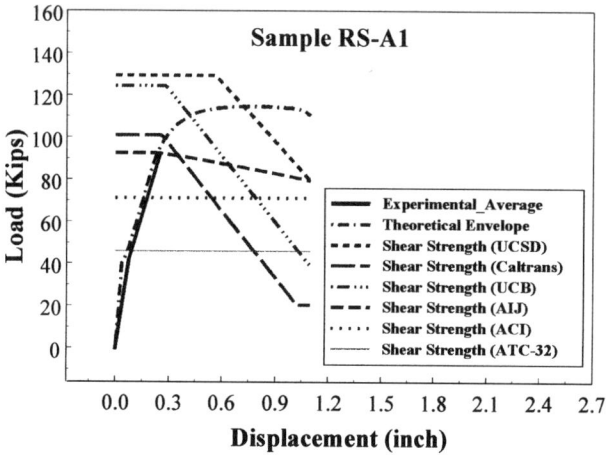

Figure 5. Experimental and Theoretical Comparison of Shear Response of Specimen RS-A1.

Retrofitted Columns

The computer code, developed in this study, was employed to evaluate the performance of tested composite-jacketed columns. The model was calibrated through a parametric study to compare between the experimental and predicted results using the two displacement models and the six different concrete confinement models that were mentioned earlier. For all tested columns, the ratio of the maximum experimental lateral load, V_{exp} to the maximum theoretical lateral load, V_{th}, and the ratio of the ultimate experimental displacement, $\Delta_{u\text{-}exp}$ to the ultimate theoretical displacement, $\Delta_{u\text{-}th}$ were calculated for the different displacement and confinement models. For circular columns, it was noted that the best fit to the experimental data was provided by Wehbe's model for displacement and Hosotani's model for FRP-confined concrete. This is indicated in Fig. 6 for specimen CS-R1. Validation of these models was also confirmed through a comparison of the experimental to the theoretical load-displacement envelopes. The experimental envelopes include both the push-pull envelope and the average envelope of the push and pull directions. This is shown in Fig. 8 for circular specimen CS-R1. However, Kowalsky's model for displacement and Hosotani's model for concrete confinement were proven to give the best fit to the experimental results for rectangular columns as shown in Fig. 7 for specimen RS-R5. These models were also validated by comparing the experimental to the theoretical load-displacement envelopes as presented in Fig. 9 for specimen RS-R5. Summaries of statistical analysis of the (V_{exp}/V_{th}) and ($\Delta_{u-exp}/\Delta_{u-th}$) ratios are shown in Table III for both circular and rectangular columns using different confinement models. It should be noted that the values listed in Table III were based on Wehbe's displacement model for circular columns and Kowalsky's displacement model for rectangular columns. It is evident that Hosotani's model for FRP-confined concrete provided the best fit to the data of all jacketed columns. Conclusively, it is important to note the success of the developed numerical analysis at predicting the behavior of squat ductile composite-jacketed bridge columns, and hence, the validity of that modeling to be used for development of retrofit design guidelines for different composite-jacket systems. Based on the statistical lower bounds in Table III, the composite jacket can be designed to retrofit circular bridge columns using the developed numerical modeling without applying any reduction factor to the predicted values. However, for rectangular columns, the developed numerical modeling can be used in retrofit design after reducing the predicted maximum load by about 10% and the ultimate displacement by about 23%.

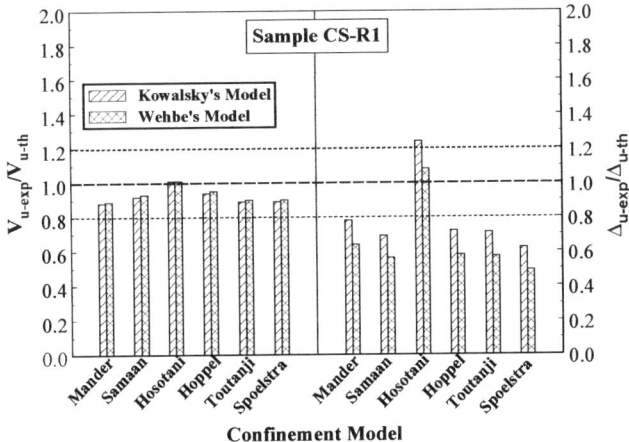

Figure 6. Experimental and Theoretical Comparison of Flexural Response of Specimen CS-R1.

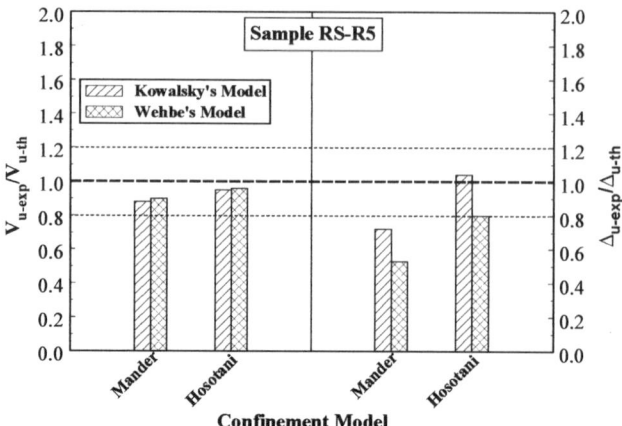

Figure 7. Experimental and Theoretical Comparison of Flexural Response of Specimen RS-R5.

TABLE III. STATISTICAL ANALYSIS FOR CIRCULAR AND RECTANGULAR COMPOSITE-JACKETED COLUMNS

Column Section	Confinement Model	(V_{exp}/V_{th})				($\Delta_{u-exp}/\Delta_{u-th}$)			
		m	σ	m - σ	m - 2σ	m	σ	m - σ	m - 2σ
Circular	Mander et al. [13]	0.898	0.008	0.891	0.883	0.636	0.002	0.635	0.633
	Samaan et al. [15]	0.943	0.015	0.927	0.912	0.593	0.047	0.546	0.498
	Hosotani et al. [10]	1.021	0.013	1.007	0.994	1.107	0.035	1.073	1.038
	Hoppel et al. [9]	0.963	0.018	0.945	0.928	0.613	0.048	0.566	0.518
	Toutanji [17]	0.908	0.010	0.898	0.888	0.566	0.008	0.558	0.551
	Spoelstra et al.	0.914	0.017	0.896	0.879	0.522	0.044	0.479	0.435
Rectangular	Mander et al. [13]	0.862	0.015	0.847	0.832	0.671	0.093	0.577	0.484
	Hosotani et al. [10]	0.934	0.013	0.921	0.908	1.009	0.119	0.890	0.771

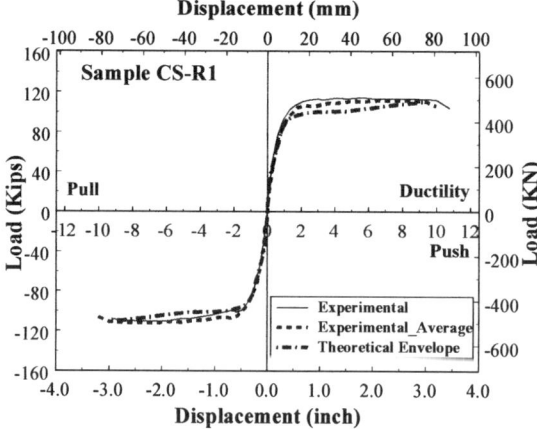

Figure 8. Load-Displacement Envelope for Specimen CS-R1.

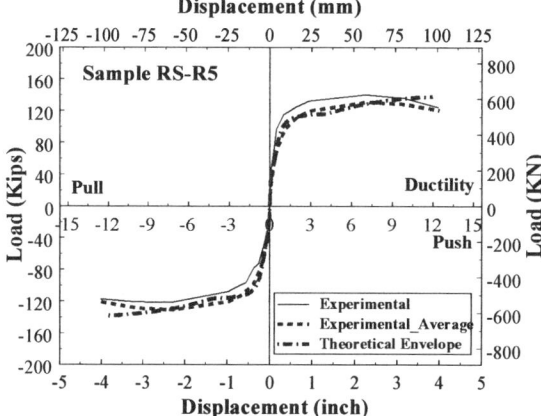

Figure 9. Load-Displacement Envelope for Specimen RS-R5.

CONCLUSIONS

A numerical analysis, compiled from a number of existing models, has been used successfully to predict the performance of squat reinforced concrete bridge columns. For shear-deficient columns, UCSD shear strength model provided the best fit to all available experimental data, and hence, can be used to predict the maximum shear force for as-built columns. To evaluate the behavior of composite-jacketed columns, it was evident that Wehbe's model for displacement calculation is most suitable for circular columns, while Kowalsky's model is more appropriate for rectangular columns. The FRP-confined concrete model that was developed by Hosotani et al. has been proven to be convenient for both circular and rectangular jacketed columns.

REFERENCES

1. ACI Committee 318, "Building Code Requirements for Reinforced Concrete and Commentary," ACI318-95 and 318R-95, Farmington Hills, MI., revised 1995.

2. AIJ Structural Committee, "Design Guidelines for Earthquake Resistant Reinforced Concrete Buildings Based on Ultimate Strength Concept (in Japanese)," Architectural Institute of Japan, 1988, 337pp.

3. Aschheim, M., Moehle, J. P., and Werner, S.D., "Deformability of Concrete Columns," Project Report under Contract No. 59Q122, California Department of Transportation, Division of Structures, Sacramento, California, June 1992.

4. ATC-32, "Seismic Design Recommendations for Bridges," Applied Technology Council, Redwood City, California, May 1995.

5. Caltrans, "Bridge Design Specifications," Sacramento, California, 1993.

6. Caltrans Memo to Designers 20-4 Attachment B, "Design/Detail Guidelines," August 1996.

7. Haroun, M.A., Feng, M.Q., "Lap Splice and Shear Enhancements of Composite-Jacketed Bridge Columns," Proceedings of the 13th US-Japan Workshop on Bridge Engineering, Tsukuba Science City, Japan, October 1997.

8. Haroun, M.A., Feng, M.Q., Bhatia, H., and Sultan, M., "Cyclic Qualification Testing of Jacketed Bridge Columns in Flexure and Shear," Proceedings of the 16th International Modal Analysis Conference, Santa Barbara, California, February 1998.

9. Hoppel, C.R., Bogetti, T.A., Gillespie, J.W.Jr., Howie, I., and Karbhari, V.M., " Analysis of a Concrete Cylinder with a composite hoop wrap," Proc., 1994 ASCE Mat. Engrg. Conf., ASCE, New York, N.Y., pp. 191-198.

10. Hosotani, M., Kawashima, K., and Hoshikuma, Jun-ichi, "A Stress-Strain Model for Concrete Cylinders Confined by Carbon Fiber Sheets (in Japanese)," Report No. TIT/EERG 98-2, Tokyo Institute of Technology, Tokyo, Japan, 1998, 55pp.

11. Joint ASCE-ACI Task Committee 426, "The Shear Strength of Reinforced Concrete Members," Journal of the Structural Division, ASCE, Vol. 99, No. ST6, June 1973, pp. 1091-1187.

12. Kowalsky, M.J., Priestley, M.J.N., and Seible, F., "Shear and Flexural Behavior of Lightweight Concrete Bridge Columns in Seismic Regions," ACI Structural Journal, Vol. 96, No. 1, January-February 1999, pp. 136-148.

13. Mander, J.B., Priestley, M.J.N., and Park, R., "Theoretical Stress-Strain Model for Confined Concrete," Journal of the Structural Division, ASCE, Vol. 114, No. 8, August 1988, pp. 1804-1826.

14. Priestley, M.J.N., Seible, F., and Calvi G.M., "Seismic Design and Retrofit of Bridges," John Wiley & Sons, Inc. New York, 1996.

15. Samaan, M., Mirmiran, A., and Shahawy, M. "Model of Concrete Confined by Fiber Composites," Journal of Structural Engineering, ASCE, Vol. 124, No. 9, Sep. 1998, pp. 1025-1031.

16. Spoelstra, M.R., Monti, G., "FRP-Confined Concrete Model," Journal of Composites for Construction, ASCE, Vol. 3, No. 3, August 1999, pp. 143-150.

17. Toutanji, H., "Stress-Strain Relationship of Concrete Cylinders Confined with FRP Composites," ACI Materials Journal, Vol. 96, No. 3, May-June 1999, pp. 397-404.

18. Wehbe, N.I., Saiidi, M.S., and Sanders, D.H., "Seismic Performance of Rectangular Bridge Columns with Moderate Confinement," ACI Structural Journal, Vol. 96, No. 2, March-April 1999, pp. 248-258.

Repair and Upgrade of R/C Two-Way Slab with Carbon/Epoxy Laminates

A. MOSALLAM, T. LANCEY, J. KREINER, M. HAROUN
and H. ELSANADEDY

ABSTRACT

This paper presents results of both analytical and full-scale investigation on the ultimate response of unreinforced concrete slabs repaired and retrofitted with fiber reinforced polymer (FRP) composite strips. Unlike the majority of the previous research work, a uniformly distributed pressure was used in loading the large-scale slab specimens. In this program, three 105" (8 ¾ ft) x 105" (8 ¾ ft) x 3" concrete slabs were tested. No steel reinforcements were used in fabricating the test specimens, except for a light steel fabric layer to avoid crack during transporting the slab specimens from the forms to the testing rig. The average 28-day concrete compressive strength for all specimens was about 3,000 psi. Five concrete cylinders were tested after seven days and at the same day of each test to determine this average. The load was applied using a specially designed pressure fixture. In this loading regime, the uniform pressure is applied to the bottom surface of the slab (top surface is in tension) via a high-pressure water bag. Load/deflection, and stress/strain curves were generated and modes of failures were recorded. Test results indicated that the load carrying capacity of the cracked slab specimen was increased about 400% by adding the external carbon/epoxy laminates. In all the tests, no catastrophic failure was observed. The paper also presents a good agreement between the experimental and theoretical results were obtained. This indicates that effectiveness of the proposed FEM in predicting the structural performance of slabs retrofitted with

---

Ayman Mosallam, Timothy Lancey, and Jesa Kreiner, California State University, Fullerton, Fullerton, CA 92834, USA
Medhat Haroun, and Hussein Elsanadedy, University of California, Irvine, Irvine, CA 92697, USA

polymer composites. This model can also be extended to develop FEM-based retrofit design guidelines for such slabs.

INTRODUCTION

For the past few years, several major research projects were launched to investigate the feasibility of using composites in both seismic and corrosion repair of structural systems made of reinforced concrete, steel, and wood materials. The majority of these programs were sponsored by the state and federal government as joint programs with the industry. The overwhelming experimental and analytical results have encouraged the civil engineers and the construction industry to consider polymer composites as an alternative construction material and system. One of the successful applications of polymer composites, is columns seismic repair and rehab. This application has been extended to cover not only the ductility enhancements, but also structural upgrading of both the flexural and axial stiffness and strength of reinforced concrete columns (Haroun *et al.* (1998)). In addition, similar applications included repair and rehab of reinforced and unreinforced bearing and shear walls, slabs, and beams. New application in seismic repair of reinforced concrete beam-column joints was recently investigated by Mosallam *et al.* (1999).

EXPERIMENTAL PROGRAM

In this program, three 8 ¾ ft (2.67 m) x 8 ¾ ft (2.67 m) x 3" (76.2 mm) concrete slabs were tested. No steel reinforcements was used in fabricating the test specimens, except for a light steel fabric layer to avoid crack during transporting the slab specimens from the forms to the testing rig. The average 28-day concrete compressive strength for all specimens was about 3,000 psi (20.67 MPa). Five concrete cylinders were tested after seven days and at the same day of each test to determine this average. The load was applied using a specially designed pressure fixture. In this loading regime, the uniform pressure is applied to the bottom surface of the slab *(top surface is in tension)* via a *high-pressure* water bag as shown in Figure (1).

At each test, the same rate of the water pressure was used. Strain gages were bonded to concrete surfaces, between layers and at the outer surfaces of the composite laminates to measure axial strains in different directions. Deflections at different critical locations were captured by five linear variable differential transducers (LVDT's) mounted on a lightweight rigid aluminum frame. The data were collected automatically via a computerized data acquisition system. The hydraulic pressure was increased gradually and crack development and propagation were continuously monitored.

Test Description: Two control slab specimens were tested initially. The first control slab specimen was tested to failure. The purpose of this test was to understand the ultimate behavior of the *unreinforced* slab and to determine crack pattern, propagation and size. The second control specimens was subjected to the same loading regime. In order to reuse this specimen in the repair program, the test was stopped at about 77% of the ultimate pressure observed in the control specimen [77% P_u= 1.23 psi (8.48 kPa)]. The partially damaged (or pre-cracked) specimen was then repaired using a combination of low-viscosity epoxy and FiberBond® carbon/epoxy laminates. The forth test was performed on an undamaged concrete slab specimen strengthened by identical carbon/epoxy laminates used for the repair test. The carbon/epoxy laminates were applied to the top surface (tension side) of the slab. A total of two strips of FiberBond® carbon/epoxy strips spaced at 18 inches (0.46 m) were bonded to the top (tension side) of the slab. The fiber architecture and lamination schedule were designed and applied such that they breathing zones between the carbon/epoxy strips are allowed (see Figure (1)). The same procedure was used in retrofitting "undamaged", unreinforced slab specimen. The lamination and geometry was the same for both repaired and retrofitted slabs.

Figure (1): Lamination and Breathing Zones of Repaired Slab Specimen

TEST RESULTS

Control Slab Specimens: As the load increased, diagonal cracks were observed. Both crack propagation and crack size continued to increase. For the first control specimen, the load was applied up to the failure pressure load of 1.51 psi (10.41 kPa), while for the second specimen, the test was halted at a load of 1.23 psi (8.48 kPa). This load is equivalent to 77% of the control specimen ultimate failure load.

Repaired Slab Specimen: As it was mentioned earlier, wide composite strips were used to repair the pre-cracked slab specimen in two orthogonal directions. The spacing between the strips was designed with maximum spacing allowed to avoid shear lag, and two-way shear of the unreinforced panels. This lamination design is practical for slabs in providing *"breathing zones"* and to avoid the degradation of the bondline by the water migrating through the top surface of the slab during service life. It should also be noticed that in laminating the strips, laminates were *staggered*. This design is preferred for several reasons, including i) to enhance the load distribution between the different laminates in different directions, ii) to maximize the energy dissipation at failure, and iii) to avoid the sudden strength degradation in case of bondline failure at certain location of the slab.

As the pressure load increased, surface cracks were observed and were concentrated in the unreinforced grid areas between the composite strips. During the course of the test, crack propagation and crack size were monitored, marked and recorded. Neither bondline, nor tensile failure of the composite laminates was observed. The ultimate load of failure was due to a shear failure of the left middle unreinforced concrete area near the support as shown in Figure (2). The ultimate pressure load of the repair specimen was 7.07 psi (48.75 kPa) which translates to a total ultimate load capacity of 77.95 kips (346.71 kN). *Thus, the use of composites did not only succeed in restoring the strength of the pre-cracked slab, but also in upgrading its strength capacity up to 475%.* In addition, the repaired specimen was able to deflect more than 2.38 inches (60.45 mm) before failure giving an ample visual warning before failure. The ultimate strain at failure was 1.27%. However, it should be noted that this was not the laminate failure strain, since the failure occurs to the concrete. Hence, the strain at failure is expected to by higher than the measured strain value of 1.27%. Figure (7) shows the load/deflection curves for all slab specimens.

Retrofitted Slab Specimen: Similar behavior was observed for the retrofitted slab. Again, the development of the surface cracks was observed at a relatively higher load level. The apparent surface cracks were only observed in the unreinforced concrete areas. The local mode of failure occurred at a pressure load of about 7.48 psi (51.57 kPa) which translates to a total ultimate load of 82.47 kips (366.81 kN). Again, the local mode of failure was in the form of a localized shear damage at the slab edge near the unreinforced corner. The corresponding mid-span deflection at the center was 2.36" (60.00 mm) while the failure strain was 1.31%. Due to the fact that the composite did not fail in this test, this strain is expected to be higher had the failure was due to tensile failure of the composite strips. Near the maximum load, a cohesive failure occurred to a very small edge portion of the one of the

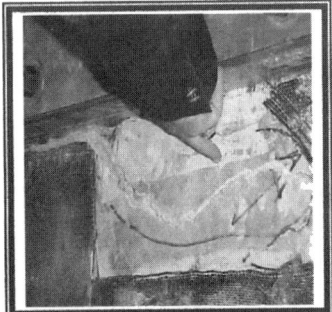

Figure (2): Ultimate Failure of the Repaired Specimen

composite strips This can be attributed to a poor surface preparation at this specific location. However, it is believed that this localized failure was not the determining factor of the strength of the retrofit system. Figure (3) shows a sample of the experimental mid-span stress/strain behavior at the retrofitted slab specimen.

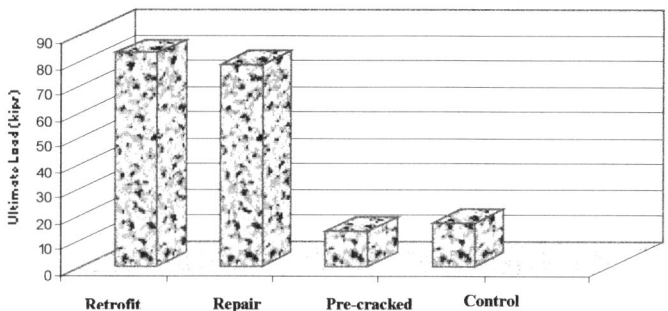

Figure (3): Strength Capacity Comparison of the Different Concrete Slab/Wall Systems

NUMERICAL MODELING

A general-purpose computer program, MARC, was used to analyze both control and retrofitted slabs by the Finite Element Method (FEM). The program has four comprehensive libraries, making it suitable for a wide range of applications. These libraries contain structural procedure, materials, elements, and program functions.

This program can be used to perform linear or nonlinear stress analysis in the static and dynamic regimes. This includes non-linearities due to material behavior, large deformation, or boundary conditions. One, two, or three-dimensional modeling is available using a variety of elements. These elements include trusses, beams, shells, and solids. MARC has another interactive program, MENTAT that is used to prepare and process data for use with the FEM. A summary description of FEM model, the elements and material libraries is discussed as follows.

Element Library: To model the slabs for out-of-plane loading, a 4-node, layered thick shell element was utilized. Three displacements and three rotation components were assigned as the element nodal degrees of freedom in the global coordinate system. Bilinear interpolation was used for the coordinates, displacements and rotations. The membrane strains were obtained from the displacement field, and the curvatures from the rotation field. The transverse shear strains were calculated at the middle of the edges and interpolated to the integration points.

Material Library: in the control slab model, each element has ten layers of concrete. however, for the retrofitted slab model, additional layers were added to the element in order to represent the composite laminates. the material behavior was integrated through the thickness of the element using the trapezoidal rule. Concrete modeling in compression was based on theory of work hardening plasticity with linear Mohr-Coulomb yield surface. the isotropic hardening rule was employed in modeling the concrete. In this case, it assumes that the center of the yield surface remains stationary in the stress space, but that the size of the yield surface expands, due to work hardening. Concrete under tensile stresses was modeled as a low-tension material. The concrete is considered cracked once the maximum principal tensile stress exceeds its direct tensile strength. This value was taken as $0.1f'_c$. after concrete cracks, linear tension softening was included. As a result, the tensile stress in the direction of maximum principal stress does not approach immediately to zero; instead the material softens until there is no stress across the crack as shown in Figure (4). A value of 0.1 was assigned for the shear retention factor. For the polymer composite system, linear elastic, orthotropic material model was used. The model parameters were based upon experimental data of coupon specimens (Mosallam (1999)).

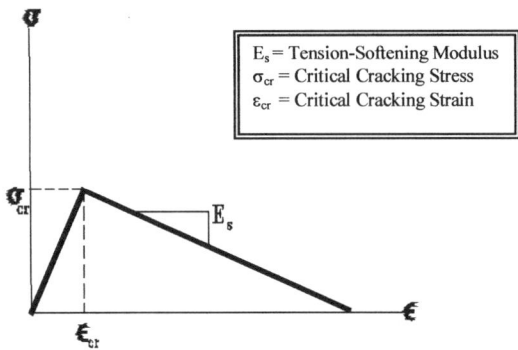

Figure (4): Uniaxial Stress-Strain Diagram for Concrete in Tension.

Solution Strategy: to solve the system of nonlinear equations incrementally, five solution schemes are available in the program. These schemes are: the full Newton-Raphson Method, the modified Newton-Raphson Method, the strain correction method, the secant method, and the direct substitution method. Of these five solution schemes, the modified Newton-Raphson method was selected for analysis used in this study. This method is effective for large-scale, mildly nonlinear problem. The program default convergence criterion was used the analysis It is based on the magnitude of the maximum residual load compared to the maximum reaction force. This method is appropriate since the residual forces are used to measure the out-of-equilibrium force, which should be minimized.

Finite Element Mesh: The finite element mesh for the retrofitted slab specimen is shown in Figure (5). The boundary conditions for both control and retrofitted specimens were simply supported at the four sides with restrained displacements and free rotations.

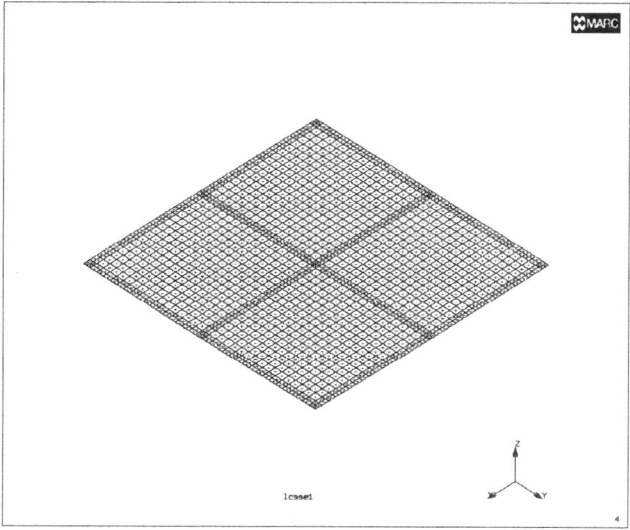

Figure (5): Finite Element Mesh for Retrofitted Slab.

Finite Element Results: The delflection conturs of the retrofitted slab is shown in Figure (6). The load-displacement envelopes generated from the FEM analysis are shown in Figure (7). For the control slab, the displacement increased as the load increases. The maximum displacement of 2.25" (57.15 mm) was calculated at a maximum load of 17.36 kips (77.21 kN), after which no convergence was possible. As shown in Figure (7), a slight increase in the initial stiffness of the retrofitted slab was observed as compared to the control specimen. However, the calculated ultimate load increased significantly by about 590%. In addition, the computed displacement at the maximum load increased by about 42%. A comparison between the experimental and theoretical results is shown in Figure (8). It should be noted that identical finite element analysis results values for both retrofitted and repaired slabs were used in constructing Figure (8). In this figure, the comparison is presented in terms of the ratio of the maximum experimental load P_{exp} to the maximum theoretical load P_{th}, and the ratio between the ultimate experimental displacement Δ_{exp} to the ultimate theoretical displacement Δ_{th}. As shown in Figure (7), a good agreement between the experimental and theoretical results was obtained. This indicates that effectiveness of the proposed FEM in predicting the structural performance of slabs retrofitted with polymer composites. This model can also be extended to develop FEM-based retrofit design guidelines for such slabs.

Ultimate Strength Design Prediction: Appendix (A) presents a simplified design approach to predict the ultimate load of the retrofitted slab described in this paper. The design procedure followed the ACI 318 code requirements for both shear and bending.

CONCLUSIONS

Based on the full-scale test results, it is concluded that the use of carbon/epoxy strips has achieved remarkable results in both repair and retrofit applications. The proposed FEM can be used as an effective tool to predict the performance of the slab retrofitted with polymer composites. A summary of the results is shown in Table (1).

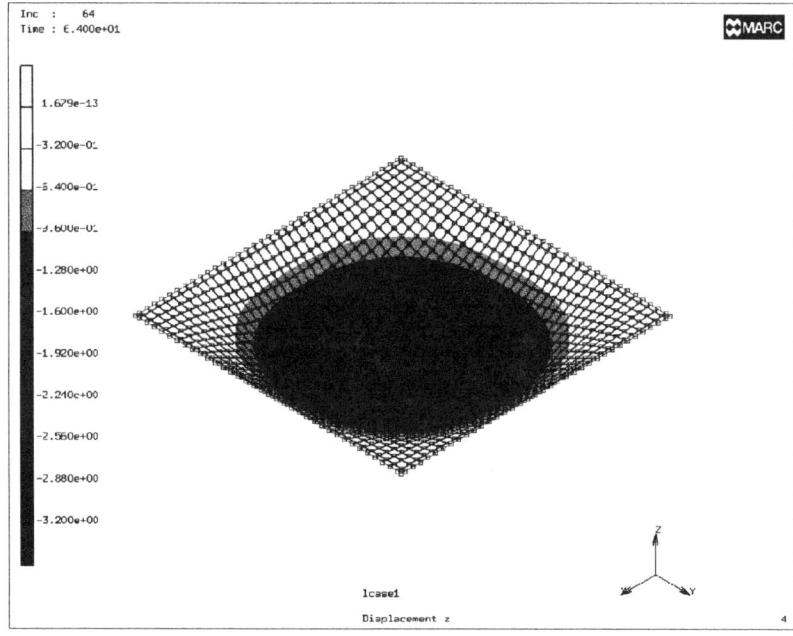

Figure (6): Deflection Contours for Retrofitted Slab

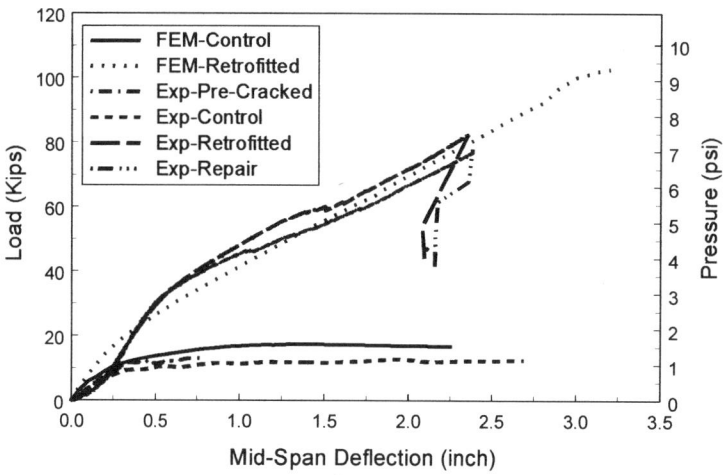

Figure (8): Comparison between Experimental and Numerical Results

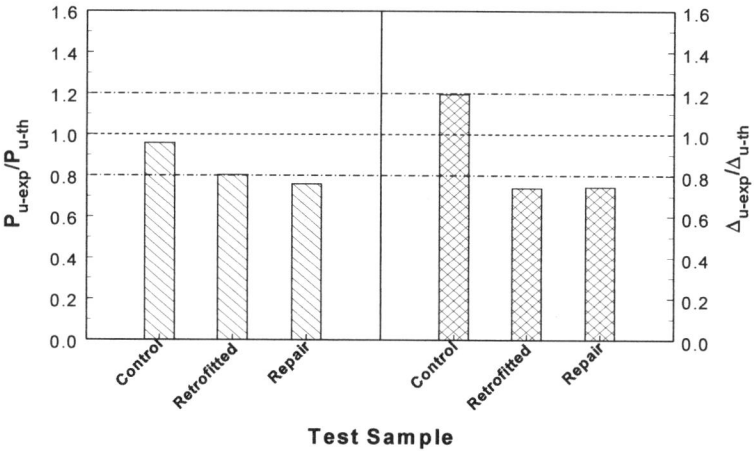

Figure (8): Comparison between the FEM and Experimental Results

Table (2): Summary Results of the Concrete Slab/Wall Repair & Retrofit Program

Specimen Description	Ultimate Pressure psi (kPa)	Ultimate Load Kips (kN)	Deflection @ Ultimate, in (mm)	Strain @ Ultimate (%)
Retrofit	7.48 (51.57)	82.47 (366.81)	2.36 (60.00)	1.31%
Repaired	7.07 (48.75)	77.95 (346.71)	2.38 (60.45)	1.27%
Pre-cracked (Repairable)	1.23 (8.48)	13.56 (60.32)	1.13 (28.70)	------
Control (Damaged)	1.51 (10.41)	16.65 (74.05)	2.69 (68.33)	------

ACKNOWLEDGMENT

The research project described in this paper is a part of the ICBO/Caltrans Structural Evaluation of FiberBond© System under two research contracts with Edge Structural Composites (ESC), LLC conducted at the SRRS Center of California State University, Fullerton (CSUF). This program is being conducted simultaneously at CSUF, University of California, Irvine, and the Aerospace Corporation. the authors would like to acknowledge the SUPPORT OF Mr. Karl Gillette and Mr. R. Martin of ECS. The effort of Mr. James Kiech in operating the tests is highly appreciated

7. REFERENCES

- Haroun, M.A.; Feng, M.Q.; Elsanadedy, H. (1998). "Experimental Study on Reinforced Concrete Bridge Columns Retrofitted using Advanced Composite Materials," A Technical Report Submitted to Caltrans, University of California, Irvine, December.
- Mosallam, A. S., Chakrabarti, P.R., and Lau, E. K. (1999) *"Concrete Connections"* Civil Engineering Magazine, Volume 69, January, pp. 42-45.
- Mosallam, A. S. (1999). "FiberBond Structural Evaluation: A Fast Track Program," A Final Report submitted to Edge Structural Composites, California State University, Fullerton, November, 50 ps.

APPENDIX (A): Design Example:

Data:
b = 12 in
$f_c' = 3000$ psi
$f_J = 123.83$ ksi
$E_J = 12000$ ksi
$t_J = 0.044$ in

Calculation of Equivalent Jacket Thickness:

$A_J = 3*18*0.044 = 2.376 \text{ in}^2$
$t_{eq} = 2.376/102 = 0.0233 \text{ in}$

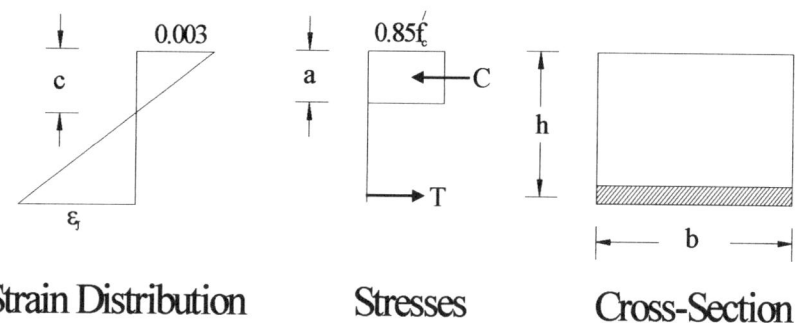

Strain Distribution   **Stresses**   **Cross-Section**

(1) Flexural Failure Mode:

$h = 3 + 0.0233/2 = 3.0116 \text{ in}$
$a = \beta_1 c = 0.85c$
$C = 0.85 f_c' ab = 0.85*3*a*12 = 30.6a$
$T = A_J * f_J = A_J * E_J * \varepsilon_J = 3355.2 * \varepsilon$
From similar triangles,
$\varepsilon_J = 0.003*(h/c - 1) = 0.00768/a - 0.003$
$T = 25.766/a - 10.0656$
$C = T$
$30.6a = 25.766/a - 10.0656$
$a = 0.768 \text{ in}$

check:
$\varepsilon_J = 0.007$
$\varepsilon_{Ju} = f_{Ju}/E_J = 0.0103 > \varepsilon_J \text{ OK}$

$M_u = 0.85 f_c' ab(h - a/2) = A_J * f_J * (h - a/2)$
$M_u = 61.735 \text{ kip.in/ft}$

$W_u = 8 * M_u/L^2 = 0.04747 \text{ Kip/in/ft} = 0.00396 \text{ ksi}$
$W_u \text{ (for Slab)} = 2.0 * 0.00396 = 0.00791 \text{ ksi}$

$P_u \text{ (for Slab)} = 0.00791 * 102 * 102 = 82.3 \text{ kips}$

(2) Shear Failure Mode:

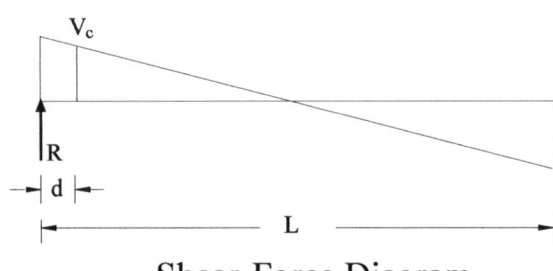

Shear-Force Diagram

Using ACI approximate method: (Section 11.3.1.1)

$$V_c = 2\sqrt{f'_c}\,bd = 2\sqrt{3000}(102)(3)/1000 = 33.52 \text{ kips}$$

$$\text{Reaction} = \frac{V_c * (102/2)}{(102/2 - 3)} = 35.615 \text{ kips}$$

$$w_u = \frac{2.0 * \text{Reaction}}{L} = 0.698 \text{ kip/in (Line load)}$$

$w_u = 0.698/102 = 0.00685$ ksi (Uniform pressure)
w_u (for Slab) $= 2.0 * 0.00685 = 0.0137$ ksi
P_u (for Slab) $= 0.0137 * 102 * 102 = 142.46$ kips

From the two failure modes, we find that flexural failure by concrete crushing occurs before shear failure occurs, and therefore:

P_u (theoretical) = 82.3 kips
P_u (experimental) = 82.47 kips

P_u (theoretical)/ P_u (experimental) = 0.998

Strengthening of Full-Scale Reinforced Concrete Beams Using FRP Laminates and Monitoring with Fiber Optic Strain Gauges

D. D. MCCURRY, JR. and D. KACHLAKEV

ABSTRACT

Four full-scale reinforced concrete beams were replicated from an existing bridge. The original beams in Horsetail Creek Bridge were substantially deficient in shear strength, particularly for projected increase of traffic loads. Of the four replicate beams, one served as a control and the remaining three were implemented with varying configurations of carbon FRP (CFRP) and glass FRP (GFRP) composites to simulate the retrofit of the existing structure. CFRP unidirectional sheets were placed to increase flexural capacity and GFRP unidirectional sheets were utilized to mitigate shear failure. Four-point bending tests were conducted. Load, deflection and strain data were collected. Fiber optic gauges were utilized in high flexural and shear regions and conventional resistive gauges were placed in 18 locations to provide behavioral understanding of the composite material strengthening. Fiber optic readings were compared to conventional gauges.

Results from this study show that the use of FRP composites for structural strengthening provides significant static capacity increases approximately 150% when compared to unstrengthened sections. Load at first crack and post cracking stiffness of all beams was increased primarily due to flexural CFRP. Test results suggest that beams retrofit with both the designed GFRP and CFRP should well exceed the static demand of 658 kN-m sustaining up to 868 kN-m applied moment. The addition of GFRP alone for shear was sufficient to offset the lack of steel stirrups and allow conventional RC beam failure by yielding of the tension steel. This allowed ultimate deflections to be 200% higher than the pre-existing shear deficient beam. If bridge beams were retrofit with only the designed CFRP failure would still result from diagonal tension cracks, albeit at a 31% greater load. Beams retrofit with only the designed shear GFRP would fail in flexure at the midspan at an equivalent 31% gain over the control specimen, failing mechanism in this case being yielding of the tension steel. Successful monitoring of strain using fiber optics was achieved. However, careful planning tempered by engineering judgement is necessary as the location and gauge length of the fiber optic gauge will determine the usefulness of the collected data.

David D. McCurry, Jr., Design Engineer, CH2M HILL, 2300 Northwest Walnut Boulevard, Corvallis, Oregon 97330-3538.
Damian Kachlakev, Ph.D., Assistant Professor, Department of Civil, Construction and Environmental Engineering, Oregon State University, 202 Apperson Hall, Corvallis, Oregon 97331-2302.

SIGNIFICANCE AND SCOPE OF RESEARCH

Upwards of 40 percent of the bridges in both the United States and Canada are structurally deficient [1,2]. Structural elements composed of concrete and reinforcing steel are frequently determined inadequate to sustain current or new load levels imposed on bridges. Recently, a new method of retrofit has been studied that uses the addition of externally bonded *fiber reinforced polymers* (FRPs) to increase load capacity.

One example, Horsetail Creek Bridge was found to be without shear reinforcing. Horsetail Creek Bridge (HCB), noted to be a historic structure, needed a retrofit scheme that maintained the original appearance of the bridge. Most traffic crossing the bridge is tourist related. Tour buses often stop on the bridge to view the adjacent falls. Since HCB beams were built without steel stirrups and were also deficient in flexure, FRPs were selected to strengthen the beams. To verify the strengthening scheme and better understand FRPs full-scale replicate reinforced concrete beams were fabricated in Oregon State University laboratories. Experimental beams were replicated to match geometry and ultimate strength of the existing beams.

This study examines the increased load carrying capacity and bending characteristics as the result of FRP added to inadequate RC beams by examining deflection and strain as a function of load. Fiber optic data complimented conventional resistive strain gauges and provided a measure of comparing the results from each. Fiber optic data was also installed on the actual bridge to monitor static, dynamic and long-term load response. Data has already been collected and will be compared to finite element models of the bridge (work in progress).

LITERATURE REVIEW

Fiber reinforced polymer materials present significant potential in the field of Civil Engineering. FRPs are not chronologically new materials, but have been used in the Aerospace industry for a number of decades. Limited demand and manufacturing of FRP material has traditionally resulted in high material costs. The first structures built utilizing these materials were in the late 1950's by Prof. H. Isler [3]. The first bridges to use FRP materials in the USA were in the late 1970's.

The need to prevent shear failure in older RC bridges has been so significant that numerous studies have focused on shear strengthening [4, 5, 6]. Full-scale and scaled experimental studies have shown that proper quantity and placement of FRP for shear strengthening is needed in order to ensure adequate strength [4]. Field applications of FRP for shear retrofit are not uncommon [7]. The number of FRP retrofit projects involving shear strengthening is likely to grow dramatically, since many reinforced concrete structures constructed in the first half of the twentieth century were inadequately reinforced for todays traffic demands.

Flexural behavior is better understood, in contrast to shear, torsion and combined loading. Basic, yet essential studies involving flexural bending of FRP reinforced RC beams have been reported [8, 9, 10, 11]. Load capacity increases are usually approximately 1.5 to 2.0 times the unstrengthened beam. Many of these studies do not involve full-scale specimens.

Horsetail Creek Bridge is a *smart structure* in that strains can be monitored real-time. Many applications of fiber optic monitoring are available now or likely to be in the future [12]. There are advantages to using fiber optic strain gauges such as less sensitivity to their environment than conventional resistive gauges. Fiber-optic technology is useful for structural monitoring and the technology is increasing in accuracy and reliability. At the time the tests were conducted, long gauge lengths (700 to 1000 mm) collected strain data for the experimental beams. Fiber-optic gauge lengths can be catered to the needs of each project. Strains can potentially be monitored over kilometers. However, for most experimental applications, small strains at discrete locations result in failure of structural members. This made fiber optics difficult to use for this study. Resolution of the fiber optic gauges used in this experiment did not provide reliable data in the linear region of load (pre-cracking). The technology is improving and future testing should consider a more precise sensor measurement.

EXPERIMENTAL DESCRIPTION

All four beams were constructed of the same geometry (6096 mm x 305 mm x 762 mm) and placement of steel and were closely matched to HCB beams. Concrete used was Type I, 20.7 MPa (28-day) strength with 152 mm slump. Beams were cured in a moist condition until removed from the forms 7-14 days after pouring. Three different FRP reinforcing schemes and control beam were used. A description of each beam is given in TAB. I. hereafter referred to as the Control, F-only, S-only and S&F beams for discussion.

TABLE I. EXPERIMENTAL BEAM DESCRIPTION

BEAM	DESCRIPTION
Control	Typical beam with no shear stirrups and bottom flexural steel
Flexure-only	Control w/ added flexural carbon FRP (CFRP) reinforcing
Shear-only	Control w/ added shear glass FRP (GFRP) reinforcing
Shear&Flexure	Control w/ added shear GFRP and flexural CFRP reinforcing

Material Properties

Structural properties used in the experiments are given in TAB. II. Note that the properties given for the carbon and glass FRP are composite properties and not properties of the fiber. All properties are design and not experimental.

TABLE II. DESIGN MATERIAL PROPERTIES

MATERIAL	LIMITING STRESS	LIMITING STRAIN	LIMIT STATE	ELASTIC MODULUS
Concrete	20.7 MPa	0.003	Crushing	21.5 Gpa
Steel	414 MPa	0.002	Yielding	200 GPa
Glass FRP	414 MPa	0.02	Rupture	20.7 GPa
Carbon FRP	760 MPa	0.012	Rupture	62 GPa

Data Collection

Three locations of deflection were collected to ensure an understanding of the beam deformation. Critical strains at the midspan section and two sections in high shear regions were collected as shown in FIG. 1. Gauges were placed on the concrete surface, FRP surface or inside the beam on the steel (appropriately protected before casting of the beams). Fiber optic gauges were utilized only on the FRP reinforced beams (i.e. no fiber optics were collected on the Control beam). These gauges were placed according to the geometry shown in FIG. 2.

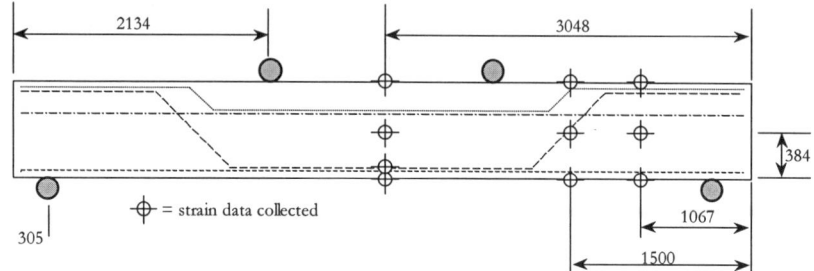

FIGURE 1. Deflection and horizontal strain data collection (internal steel reinforcing shown as dashed lines)

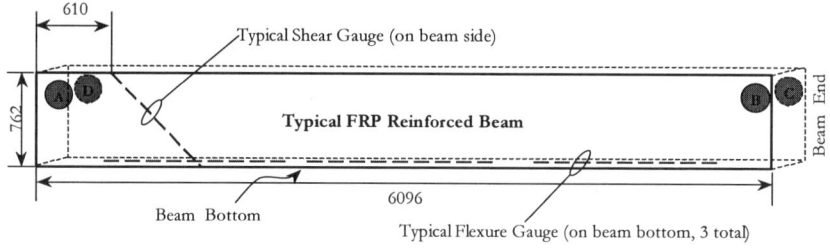

FIGURE 2. Fiber optic gauge locations

EXPERIMENTAL RESULTS

FRP reinforced beams were substantially stronger that the unreinforced Control beam. Failure modes are given in TAB. III. The fully-reinforced S&F beam did not fail (equipment limited maximum applied load), but was tested beyond the design truck loads for HCB. All reinforced beams were stronger and stiffer as shown in FIG. 3. Beam cracking was recorded to compliment the deflection and strain data. As visible in FIG. 4 the shear FRP reinforcing successfully stopped the shear cracks from forming (or at least growing). The S&F beam revealed very minimal cracking. A summary of the experimental load and deflection results is presented in TAB. IV. From this table, it is clear that the addition of FRP not only increases the sustainable load, but deflection characteristics changed. All beams were flexurally stiffer and since the Control beam was shear deficient (failing long before flexural capacity) ultimate deflections for FRP strengthened beams were increased.

TABLE III. BEAM FAILURE MODES

BEAM	FAILURE MODE
Control	Diagonal tension crack (shear failure)
F-only	Diagonal tension crack (shear failure)
S-only	Yielding of tension steel followed by crushing of compression concrete after extended deflections
S&F	Not observed. Believed to be yielding of tension steel followed by crushing of the concrete.

TABLE IV. SUMMARY OF EXPERIMENTAL LOAD AND DEFLECTION

ITEM	CONTROL VALUE	COMPARISON TO CONTROL BEAM (%)		
		F-ONLY	S-ONLY	S&F
Midspan Deflection @ 67 kN	1.18 mm	103	105	93.5
Max Observed Deflection	24 mm	124	144	104[3]
Midspan Deflection @ failure	24 mm	124	208[1]	Not Observed
Load @ failure	476 kN	145	145	Not Observed[2]
Load @ first significant cracking	78.3 kN	123	112	123
Calculated Moment Capacity	473 kN-m	188	100	188
Experimental Maximum Applied Moment	436 kN-m	145	145	200[3]

1. Extrapolated from data (i.e. device failed).
2. 152% was achieved during loading.
3. These values are at maximum applied load, not failure.

It is important to realize that the experimental gains resulting from using the FRP would not be as significant if the beam was only deficient in flexure. Even after an adjusted load configuration to increase the applied moment, the S&F beam did not show critical strain values. Deflections were visible, but the beam did not exhibit signs of failure. The load was held for a number of minutes to observe real time strains, but no indication of the beam achieving critical strains was visible. Deflections were more evident than the first configuration (third-point), but did not increase under a constant 712 kN load. This fully reinforced beam was then kept for cyclic loading to gain understanding of fatigue behavior.

Fiber Optic Strain Data

Much of the data collected from the fiber optics was ultimately not useful for behavioral interpretation. This was the result of the gauge length used being too long to extract representative strain response from the beam testing. Fiber optic gauges in the shear region collected data over a 700 mm gauge length, which included a shear crack near the beam bottom and near-zero strains near the top of the beam. Strains near the bottom of the beam and near the support are very high due to diagonal tension cracking. In contrast, strains near the top of the beam over the supports are very small. Thus, providing a gauge over this distance will only produce average strains, which are not useful for structural analysis of a concrete beam.

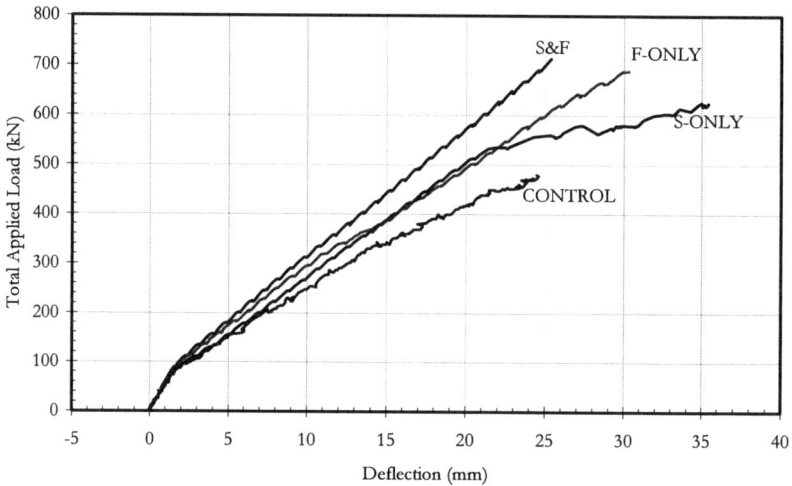

FIGURE 3. Load vs. deflection for experimental beams.

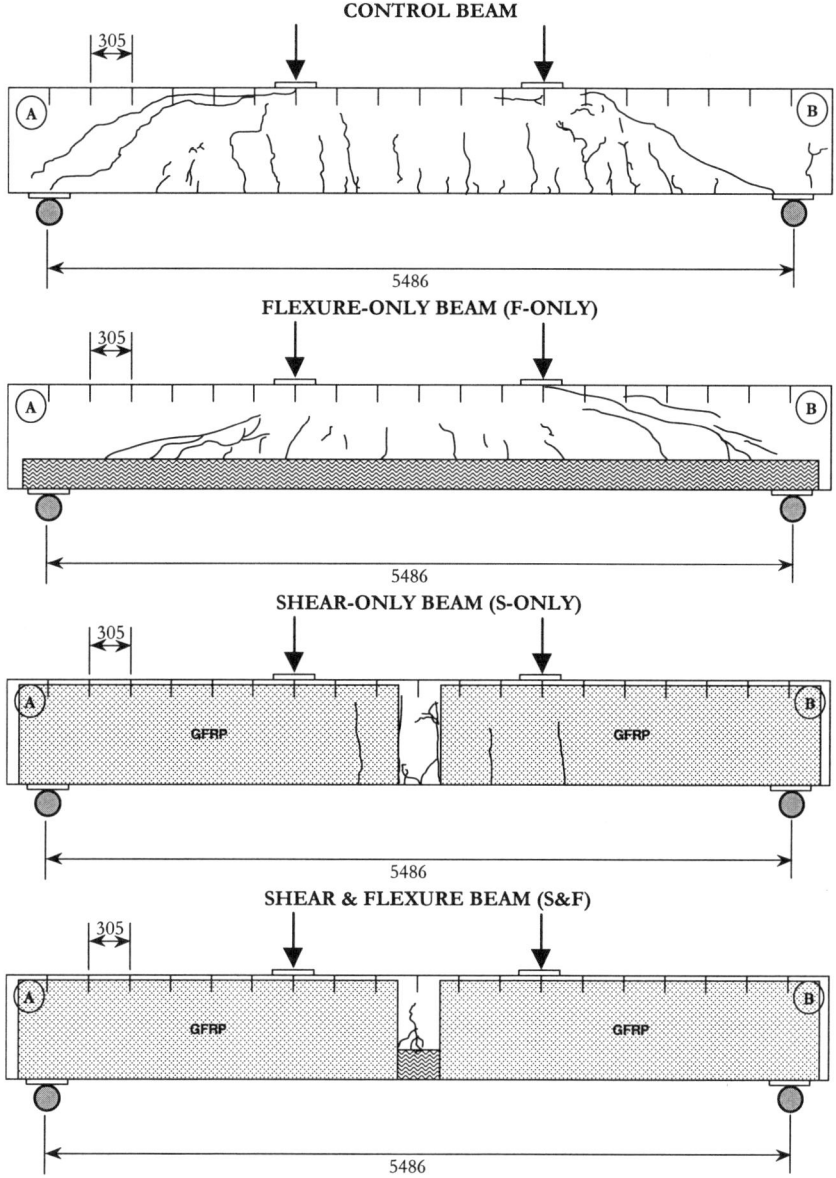

FIGURE 4. Experimental beam cracking

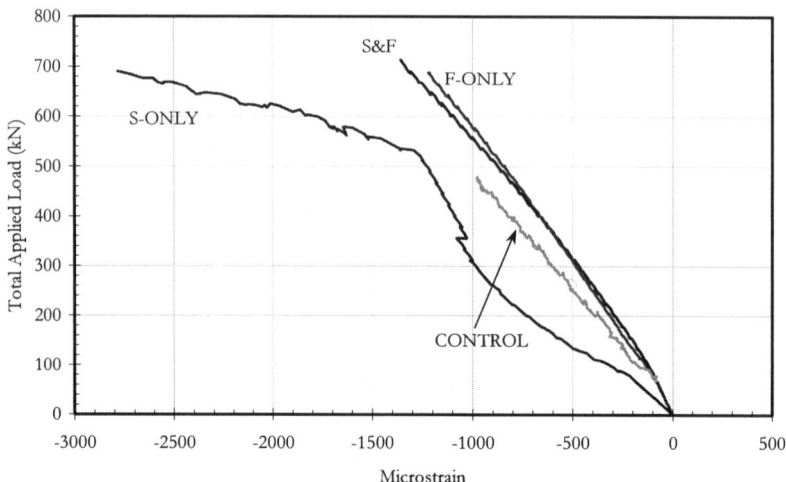

FIGURE 5. Experimental load vs. top compressive strain at midspan

Lag in FRP Strain

Nearly every calculation of the moment capacity of an FRP reinforced beam assumes that FRPs are perfectly adhered to the beam surface. Although we do understand and accept that a "perfect" bond does not exist, it would be desirable to quantify how imperfect the bond might be. Although this experimental study only involved four specimens a deviation from the plane-sections-remain-plane assumption was observed. This was evident for the F-only and S&F beam in particular as visible from the strain profiles in FIG. 6. This figure suggests that the strain observed in the FRP was near that in the main reinforcing steel. This is obviously not in agreement with common bending assumptions. This behavior is likely because the fibers were not initially taught/straight, since the FRP was applied by hand without pretensioning. A more detailed understanding of this behavior is needed to truly predict the capacity of an FRP strengthened concrete beam.

To acquire a good bond to the beam surface, proper surface preparation of the concrete beam requires a number of steps, including mechanical abrasion. This was done with experimental and HCB beams. In addition, FRP systems must be approved for use. The system used in strengthening experimental and HCB beams was an approved system, commonly used in projects worldwide. This phenomenon should be further investigated.

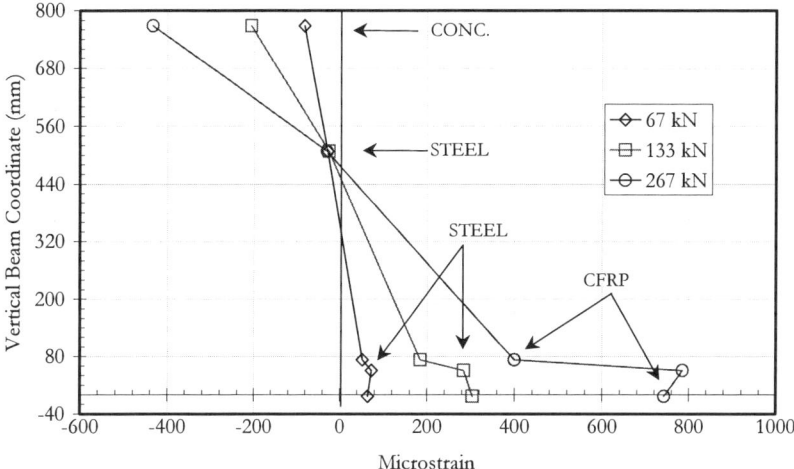

FIGURE 6. FRP horizontal strain "lag" observed from F-only beam test

CONCLUSIONS

These are observations with adequate evidence to be considered conclusive.
- ❖ FRP retrofit for structural strengthening can increase static load capacity upwards of 150% of the original beam capacity (depending on the mode of failure, geometry and material properties).
- ❖ HCB beams retrofitted only with the flexural CFRP would still result in diagonal tension failure albeit at a 31% greater load. Since the CFRP was intended to provide flexural reinforcing, it was horizontally unidirectional. The CFRP was wrapped up the sides a sufficient amount to provide resistance across the diagonal tension crack. In addition, the increased stiffness provided by the CFRP decreased the deformation and offset cracking by reducing strain in the beam.
- ❖ The addition of GFRP for shear was sufficient to offset the lack of stirrups and cause conventional RC beam failure by steel yielding at the midspan. This allowed ultimate deflections to be 200% higher than the shear deficient Control beam, which prematurely failed due to a significant diagonal tension crack.
- ❖ HCB beams retrofitted with both the GFRP for shear and CFRP for flexure should well exceed the static demand imposed by the new traffic loads of 658 kN-m sustaining up to 868 kN-m applied moment.
- ❖ Load at first crack was increased, primarily due to the added flexural CFRP, by approximately 23%. The flexural CFRP reduced the post-cracking deflections, which in turn reduced the strains and stresses in the cross section.
- ❖ Addition of flexural CFRP offset the load at yielding of tension steel beyond 33%.
- ❖ An imperfect bond was evidenced from the strain lag of the CFRP from the expected plane-sections-remain-plane assumption. Due to the extremely small

sample size of the experiment (i.e. only four specimens each with different reinforcing), further investigation is necessary to determine if this effect is of significance in structural analysis and safety.

RECOMMENDATIONS

To fully quantify the behavior of FRP reinforced beams it is recommended that statistical full-scale studies be conducted. Many studies published to-date do not involve full-scale specimens. Few publications report large sample size, full-scale testing, undoubtedly because of economic reasons. For increased confidence of beam response and to avoid future waste of FRP materials, full-scale specimens should be tested. It is also recommended that investigation be conducted into the strain lag of FRP (slightly imperfect bond observed). This phenomenon may cause unacceptable inaccuracies in calculations. In addition to experimental studies, analytical investigation should be made into the strain lag of FRP (slightly imperfect bond observed). Shear deformations of the epoxy resin can be estimated with theoretical calculations. Such studies are likely to help develop more accurate assumptions to be used in design.

REFERENCES

1. Cooper, J.D. "A New Era in Bridge Engineering Research." *2nd Workshop on Bridge Engineering Research in Progress*, NSF, Reno, Nevada, pp. 5-10.
2. Rizkalla, S., P. Labossiere. "Planning for a New Generation of Infrastructure: Structural Engineering with FRP—in Canada." *Concrete International*, Oct. 1999, pp. 25-28.
3. Meier, U., M. Deuring, H. Meier, G. Schwegler. " Strengthening of Structures with CFRP Laminates: Research and Applications in Switzerland." *Advanced Composite Materials in Bridges and Structures*, ACMBS-MCAPC, K.W. Neale and P. Labossiere, Editors; Canadian Society for Civil Engineering, 1992, 243-251.
4. Cheng, R.J.J., R. Hutchinson, S.H. Rizkalla. "Rehabilitation of Concrete Bridges for Shear Deficiency Using CFRP Sheets." *42nd International SAMPE Symposium*, May, 1997, 325-335.
5. Arduini, M., A. D'Ambrisi, A. Di Tommaso. "Shear Failure of Concrete Beams Reinforced with FRP Plates." *Infrastructure Repair Methods*, PUBL, MONTH, 199X, 123-130.
6. Al-Sulaimani, G.J., et. al. "Shear Repair for Reinforced Concrete by Fiberglass Plate Bonding." *Structural Journal*, ACI, Vol. 91, No. 3, July-August, 1994, 458-464.
7. Crasto, A.S., R.Y. Kim., et. al. "Rehabilitation of Concrete Bridge Beams with Fiber-Reinforced Composites." *42nd International SAMPE Symposium*, May, 1997, pp. 77-83.
8. GangaRao, H.V.S., P.V. Vijay. "Bending Behavior of Concrete Beams Wrapped with Carbon Fabric." *Journal of Structural Engineering*, ASCE, Vol. 124, No. 1, Jan., 1998, 3-10.
9. Rostasy, F.S., C. Hankers, E.H. Ranisch. "Strengthening of R/C- and P/C-Structures with Bonded FRP Plates." *Advanced Composite Materials in Bridges and Structures*, ACMBS-MCAPC, K.W. Neale and P. Labossiere, Editors; Canadian Society for Civil Engineering, 1992, 255-263.
10. Ritchie, P.A., et. al. "External Reinforcement of Concrete Beams Using Fiber Reinforced Plastics." *Structural Journal*, ACI, Vol. 88, No. 4, 1991, 490-496.
11. Saadatmanesh, H., M.R. Ehsani. "RC Beams Strengthened with GFRP Plates I: Experimental Study." *Journal of Structural Engineering*, ASCE, Vol. 117, No. 11, Nov., 1991, 3417-3433.
12. Udd, E. "Fiber Optic Smart Structures." *Proceedings of the **IEEE***, IEEE, Vol. 84, No. 1, Jan. 1996, pp.60-66.

Seismic Repair and Rehabilitation Using FRP Composites: A Systematic Approach

G. R. STEVENS

ABSTRACT

Quality repairs for buildings and bridges begin with understanding the importance of a systematic approach. The engineer responsible for determining which repairs are required for a structure should be familiar not only with new design and construction, but with the evaluation of distressed structures, and differing repair procedures. To determine appropriate repairs, an engineer must see the complete picture of structural behavior and not just specific points of distress.

The structural engineer should identify the client's purpose in completing all required repairs and his work objectives prior to starting work. Knowing and understanding different types of repairs is a key factor in defining expected repair performances.

REPAIR TYPES

Most repair types can be placed in one or a combination of the following categories:
- (1) Maintenance
- (2) Cosmetic
- (3) Structural
- (4) Upgrades
- (5) Modifications

Maintenance: Periodic installation of materials to slow or reduce the affects of aging and weathering. Maintenance includes:
- painting
- roofing
- caulking
- sealing cracks or concrete surfaces
- maintaining a water tight envelope.

In most structures, maintenance is required to enhance material durability.

Gene R. Stevens, S.E. Consulting Engineers, *4045 Vineland Avenue, Suite 202, Studio City, CA 91604*

Cosmetic Repairs - restoration of esthetics in structural or non-structural elements.

A few types cosmetic repairs include:
- removing stains
- repairing cracks in non-structural elements
- installing patch repairs
- replacing exterior and interior envelopes.

Structural Repairs - restoration of structural capacity and/or restoring continuity.

For concrete buildings structural repairs include:
- injecting concrete cracks with epoxy,
- patching concrete spalls and confining region with composites,
- replacing reinforcement, adding reinforcement, or adding composites in concrete,
- adding or replacing members and connectors (composites) (see Figure (1)).

Figure (1): Application of Carbon/Epoxy Composite System

Structural Upgrades - increasing the structural strength and ductility (energy absorption capacity) of members and systems.

Some types of structural upgrade include the following:
- adding members and composite reinforcement (see Figure (2)),
- adding composites,
- confining brittle materials (composites),
- adding connectors.

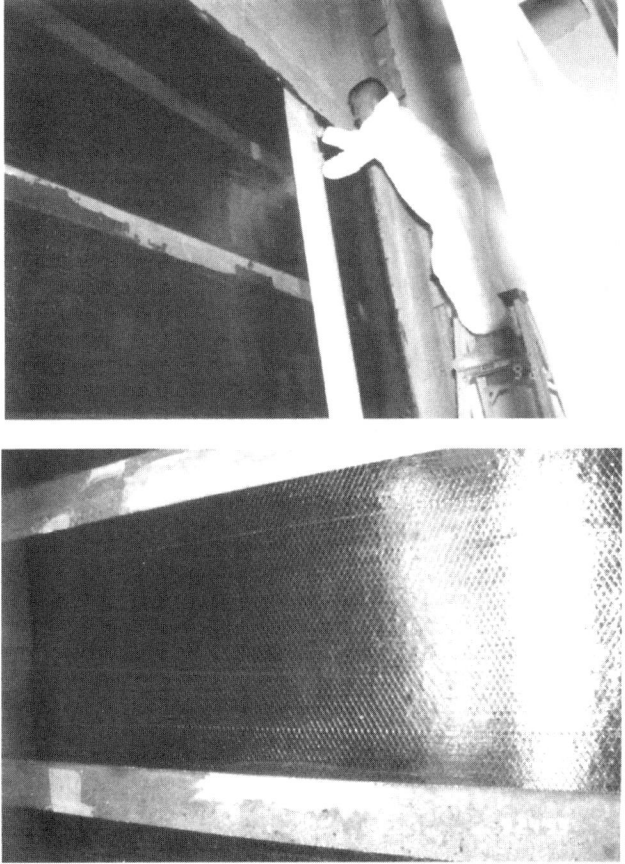

Figure (2): External Reinforcements of Structural Members Using Polymer Composite Laminates

Modifications - changing the characteristics of members, systems, or parts of structures. Modifications include:

- constructing additions
- adding new openings
- changing the strength and/or stiffness of members and systems
- changing the use or loads

SYSTEMATIC REPAIR PROCESS

To select proper repairs, a systematic repair process should include the following:
(1) reviewing existing documentation and understanding the history of the structure
(2) surveying field conditions
(3) evaluating distress to determine the most likely causes
(4) reviewing repair goals vs. structural safety and available funds
(5) selecting proper repairs (materials and procedures)
(6) Installing the correct repair

Review of Documentation: The review of documentation establishes the basis for understanding the behavior that has caused the distress and developing the appropriate repairs. This phase of the process includes:
  A. reviewing existing drawings and reports
  B. conceptualizing structural and architectural systems and existing materials
  C. defining repair goals and Code requirements
  D. initially reviewing the structure and distress
  E. listing elements to survey and test
  F. questioning observers to develop historical insight

Survey of Field Conditions: The survey of field conditions should include:
  A. viewing and documenting structural and architectural systems
  B. documenting distress as needed (Photos, Sketches etc.)
  C. Identifying potential problems (variations in distress and conditions that will likely affect repairs)
  D. conducting necessary tests

Evaluation of Distress & Determination of Cause: The engineer must determine why the existing structure has or has not functioned as intended in the initial design and why the structure is distressed in one region and not in another. I start this process by visualizing what load or displacement may have caused the viewed distress. This phase of the repair process will include the following:
  A. checking historical background - events that imparted foreseeable or unforeseeable loads or displacements such as the following:
    - *natural or unexpected soil movements*
    - *seismic events or storms*
    - *environmental conditions*
    - *overloads*
  B. checking improper design or construction errors and changes in the state-of-the-art of engineering
  C. reviewing past and present Standards and Codes
  D. estimating current strengths, serviceability, and ductility requirements as needed to determine structural deficiencies and safety

Review of Repair Goals vs. Structural Safety & Available Funds: If safety is an issue, then the probability of loads or additional movement causing partial or complete collapse of a member, system, or structure must be addressed. Most Codes mandate the prompt repair of any unsafe and potentially unsafe structural conditions. For seismic events, FEMA has life-safety guidelines. Limiting loads or uses should not be overlooked when investigating possible solutions to structural strength issues.

If funds for repairs are a problem, then structural repairs or upgrades can extend over a period of time using a priority system for repairs. Durability is the second most important issue in selecting proper conceptual repairs. Costs and in-place maintenance policies must be evaluated to determine if limited (short durability) repairs are appropriate or if only long-term repairs or upgrades need be considered. Choose the type of repair that best fits the client's needs and his pocket book then define the purpose for each repair.

Select Proper Repairs (Materials and Procedures): Once objectives of each repair are defined, then supplier's guidelines and/or data sheets are used to design preliminary repairs and select materials for cost estimating purposes. If necessary, different repair types, repair procedures, and costs are provided for review by others prior to initiating the final design and preparing repair documents.

The first step in the final design is to review existing field conditions to ensure that selected repair materials can be installed. Unless all distress has been viewed, unforeseen variations in distress and field conditions will arise during construction. The engineer should anticipate unforeseen field issues during installation and be prepared to modify repairs for these conditions. Prepare construction documents for the generally observed distress with provisions addressing unforeseen issues.

Install the Correct Repair: Viewing all field conditions is costly and impractical, since in buildings structural elements are generally covered with architectural materials. Therefore, significant construction-phase engineering will nearly always be required in all repair projects. Recommendations to ensure the correct repair is installed include:

(A) reviewing repairs in the field with all parties prior to starting any installation
(B) viewing field circumstances at previously viewed distress and installation of repairs in those regions
(C) viewing regions of unknown conditions as they are uncovered to minimize the impact of unforeseen issues and to address modifications as they arise
(D) documenting all variations in distress and field circumstances
(E) preparing field modifications for unforeseen conditions
(F) maintaining quality installation by proper inspection and structural observations

CONCLUSIONS

Providing the correct repairs for structures involves understanding repair types and completing an evaluation of the structure to define the most likely cause for the distress. Structural engineers providing repair solutions should have experience in construction, new design, evaluating distress structures, and designing and installing repairs. Knowledge of the structure and its history provide valuable insight in the selection of proper repairs. Engineers should use a systematic approach to the repair of structures. Owners or their representatives will benefit from selecting engineers that use these systematic repair approaches.

ACKNOWLEDGMENT

The construction repair work was performed by Wallock & Maggio, Inc.

Design Philosophy for Strengthening with Carbon Fiber Reinforced Polymer Composites

E. FRETT

ABSTRACT

Due to high cost of new construction and the increasing inventory of otherwise adequate concrete structures, the need for repair or strengthening of reinforced concret e structures is growing significantly in the US. In response to this need, a new structural strengthening technology has emerged. Steel-plate bonding has been a concrete-repair concept for the past twenty years. However, the high cost of labor to set the usually heavy plates, the difficulties with splicing the plates, and concerns about corrosion of the steel plates have limited the use of this technique. Because of Sika CarboDurs excellent weight -to-strength properties, recently reduced material costs, relatively unlimited material-length availability, comparably simpler installation, and immunity to corrosion, the use of Sika CarboDur for post strengthening is gaining favor over steel plate bonding and is advancing the concept of externally-bonded strengthening. Ongoing development of cost-effective production techniques for fiber reinforced polymers (Sika CarboDur) continues to increase their application.

The use of FRP in the United States lags that in Europe and Japan, and no clear rules for concrete design with FRP reinforcement exist in United States today. The American Concrete Institute (ACI) Committee 440-F is developing a document (ACI 440-F-99, 1999) which addresses design recommendations and construction techniques for the use of FRP for concrete reinforcement. The committee report will summarize the existing state-of-the-art of the industry and the fundamental design philosophy. Design with FRP reinforcement, as discussed by ACI 440-F, follows the same basic principles of equilibrium and constitutive behavior as used for conventionally reinforced concrete. However, explicit methodology for addressing several important issues, such as appropriate safety factors, low ductility of FRP composites, and anchorage/development issues, remains to be defined.

Sika Corporation, 201 Polito Avenue, Lyndhurst, NJ 07071

The objective of this paper is to emphasize and discuss several key issues related to the methodology for the design of post-strengthening of concrete structures using Sika CarboDur. The basic design philosophy presented here aligns with and expands upon the guidelines being considered by the American Concrete Institute (ACI) Committee 440-F.

Several design/philosophy issues that FRP-strengthening designers must consider are:

- Minimum required pre-strengthened strength criterion to prevent collapse if the Sika CarboDur is comprised due to uncontrollable events (fire, vandalism, impact, etc.).
- Limits on strength enhancement to maintain "ductile" behavior.
- Appropriate ϕ ("reliability") factors and limits on design strength.

Threshold Strength of Structures Prior to Strengthening

A major obstacle in the development of externally bonded strengthening concepts is associated with the risks of loss of FRP effectiveness due to uncontrollable events. The direct risk is damage due to exposure to fire; high temperatures in a fire will cause the fixing adhesive/epoxy to flow plastically causing a loss of load transfer to the FRP. Typically, the critical temperatures for the epoxy, generally related to HDT (heat-deflection temperature) or to the glass transition temperature, are in the range of 120 °F to 200 °F. Traditional fireproofing materials and systems cannot protect to such low temperatures. In recognition of the temperature risks, members to be strengthened with Sika CarboDur should posses an unstrengthened capacity that provides a positive factor of safety against collapse. The unstrengthened capacity should be the result of a ductile combination of concrete and steel, since the strengthened structure may have experienced unusual cracking prior to FRP loss due to higher than originally intended load exposures. In flexural elements, for instance, this implies not only adequate tension steel for flexure, but also well distributed steel stirrups that "hang" load in the analogous truss and keep diagonal cracks tight enough to allow appreciable shear transfer across crack faces. We propose that the unstrengthened structure be capable of resisting the service loads without yielding of the reinforcing steel and that the ultimate strength of the unstrengthened system exceed the service loads by a factor of 1.2:

$$\phi S_n \geq 1.2 (S_D + S_L + ...)$$

The above provides implied safety factors of 1.2/0.9 = 1.3 for flexure and 1.2/0.85 = 1.4 for shear. This level of "safety" will prevent yielding of the reinforcing steel. This level of "safety" is similar to that provided by the ACI guidelines for load testing: for typical ranges of dead and live loads, the ACI load test to 0.85 (1.4D+1.7L) implies a "safety factor" of ±1.3.

While this limit appears to severely restrict the applicability of the FRP strengthening technique, the following must be considered:

- Safety factors are necessary to account for the probability of the coincidence of the 1) unintended load, 2) understrength material, 3) unintended construction influences, and 4) unintended environmental influences. The entire reserve of strength implied by the "safety factor" cannot be consumed by one demand; a FRP strengthened structure, compromised by FRP loss in a fire can also be coincidentally overloaded and compromised by hidden original construction anomalies.

- A load increase from 1.2 (D+L) to 1.4D+1.7L is still significant. These limits still offer opportunities for meaningful strengthening. For example, a typical office slab with a dead load of 125 psf and an original design live load of 50 psf can be strengthened within the limit to accept a new live load of 91 psf.

- Until fire protection methods are developed to protect the Sika CarboDur system against exposure in the 150°F to 200°F range, a significant design-based protection is the fire protection rationale.

- Experience with FRP is limited; many performance issues are still incompletely tested and many environmental exposures have not yet passed the test of time. Caution is prudent.

As will be demonstrated below, other limits to ensure ductility, bond, anchorage, and strain comparability are similarly restrictive.

Minimum Steel Ratio to Ensure "Pseudo-Ductile" Behavior

Conventionally reinforced concrete members are ductile due to the presence of steel reinforcement. To ensure ductile failure of flexural members (steel yielding before concrete crushing), ACI prescribes a limit to the amount of reinforcement allowed in a member in terms of the ratio of the area of reinforcement to the area of concrete (reinforcement ratio, ρ):

$$\rho = A_s/bd \leq 0.75\ \rho_{bal}$$

Where ρ_{bal} is that reinforcement ratio which results in simultaneous concrete compression failure and steel yielding failure under flexure. The above requirement guarantees that the structure, if overloaded, will exhibit excessive deformation and thereby provide visible warning of collapse.

Sika CarboDur is not a ductile material, exhibiting nearly linear stress-strain behavior when loaded to failure in tension. However, experimental tests show that steel-reinforced concrete flexural members strengthened with Sika CarboDur can exhibit ductile behavior when loaded to failure.

Experiments have shown that the moment-rotation curves for Sika CarboDur-enhanced, reinforced concrete beams are bilinear with the portion between steel yielding and failure having an upward slope rather than being horizontal as is the case of the idealized elastic-plastic curves for members reinforced with steel only. This phenomenon, called "pseudo-ductility", is not unexpected, since conventionally reinforced concrete elements should not lose their ductility due to introduction of a non-ductile material, in this case Sika CarboDur.

The slope of the upper portion of the moment-rotation curve depends on the ratio of FRP reinforcement to steel reinforcement. As this ratio increases, the slope steepens and at the extreme it will match that of the lower portion of the curve with the member exhibiting essentially linear-elastic behavior (see the figure below).

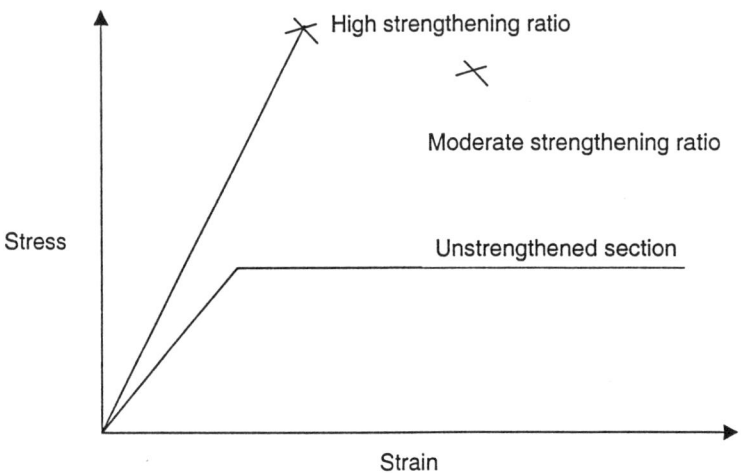

The strengthening designer must evaluate whether the strengthened member possesses sufficient ductility. One method to examine the level of ductility for a flexural member is to plot an idealized moment-rotation curve for the post-strengthened member and compare it to the idealized moment-rotation curve for the unstrengthened member if reinforced to 0.75 ρ_{bal}.

It is also important that the designer examines and understands the behavior of the strengthened member at failure. This requires not only traditional evaluations like examination of limits associated with the rupture of the steel or FRP, or the crushing of the concrete, but bond-related FRP-system failure modes as well. These failure modes are associated with sudden loss of bond, or sudden delamination due to the strain incompatibility of the reinforced concrete substrate and the FRP composite. For instance, a strengthened flexural member reinforced such that the reinforcing steel will fracture when the FRP material is stressed only to a fraction of its ultimate strength, has no ductility. This would be an unacceptable design. The designer must understand the behavior of the member prior to and after strengthening, and must ensure that the member contains enough steel to provide ductile behavior at or near ultimate loads. A method currently under consideration by ACI 440-F (1999) requires that the minimum strain in the mild steel reinforcement at failure of the strengthened section (regardless of the failure mode), be roughly 2.5 times the strain in the steel at yield. This level of over-strain may be impractical, but the concept is valid.

In addition to the statistical reliability of the material and of the manufacturing and installation processes, there may be limits on strength mobilization related to the strain compatibility of the substrate or parent member material. A basis of ACI 318 is a concrete strain limit of 0.003, to define crushing failure of the concrete. In addition, it is commonly reported that aggregate interlock in concrete becomes unreliable at concrete strains in excess of 0.006. While Sika CarboDur in some applications can overcome these strain limits, the substrate will limit the utilization of its full potential in some failure modes. Since FRP research is still in the early stages and many practical variables are still to be studied and evaluated, simple strain limits are appropriate to define empirically developed limit states. Where applicable, these should be conservatively selected to act as direct limitations applied to the section in the analysis process, without modification by ϕ factors. Currently, for a sustained load application, we propose that Sika CarboDur strains be limited to 0.0045 in flexure (medium strain away from the cracks) and that Sika CarboDur strains in shear and in compression be limited to 0.004. For example, in flexure, the medium strain limit of 0.0045 effectively limits the usable strength of the Sika CarboDur strips to approximately 40% of the ultimate strength ($f_{Lueff} = E_L \times 0.0045/k_L$, where k_L is the ratio of maximum and medium strains). Further work must be undertaken to evaluate if there should be differences in strain limitations relative to transient vs. sustained load conditions and life safety.

References

ACI 318-95, 1995. Building Code Requirements for Structural Concrete (ACI 318-95) and Commentary (ACI 318R-95).

ACI 440R-96, 1996. State-of-the-Art Report on Fiber Reinforced Plastic Reinforcement for Concrete Structures. Reported by ACI Committee 440, American Concrete Institute, Detroit, Michigan 48219.

ACI 440F-99, 1999. Guidelines for the Selection, Design, and Installation of Fiber Reinforced Polymer (FRP) Systems for Externally Strengthening Concrete Structures. Reported by ACI Committee 440, American Concrete Institute, Detroit, Michigan 48219.

Canadian Standards Association (CSA) publication A23.3-94, Design of Concrete Structures

Karbhari, V.M. and Sieble, F. 1997. Design Considerations for the Use of Fiber Reinforced Polymeric Composites in the Rehabilitation of Concrete Structures.

Kelley, P.L., Brainerd, M. L., Vatovec, M., 1997. Engineering Guidelines for the Use of Sika CarboDur (CFRP) Laminates for Structural Strengthening of Concrete Structures, Chapter 4: Design Guidelines. Sika Corporation, Lyndhurst, NJ.

Meier, U., 1995. Strengthening of Structures Using Carbon Fibre/Epoxy Composites. Construction and Building Materials, Vol. 9, No. 6, pp. 341-351.

Nanni, A., Cobb, C., 1993. Design Examples. FORCA TOW SHEET Technical Manual.

Sika 1996: Reinforced Concrete Structures with the Sika CarboDur-System. Technical Manual.

Schweizerischer Ingenieur-und Architecten-Verein, 1989: SIA 162; Betonbauten (Concrete Structures). Swiss Code SN 562 162.

Tausky, 1993. Betontragwerke mit Aussenbewehrung. Birkhäuser Verlag. ISBN 3 - 7643 - 2911 - 4.

Triantafillou, Thanasis C., Deskovic, Nikola, and Deuring, Martin 1992. Strengthening of Concrete Structures with Prestressed Fiber Reinforced Plastic Sheets. ACI Structural Journal, V. 89, No. 3, May-June 1992.

Van Gemert, D. 1992. Special Design Aspects of Adhesive Bonding Plates. ACI SP-165-2: Repair and Strengthening of Concrete Members with Adhesive Bonded Plates. pp.25-41.

Winter, George, and Nilson, H. Arthur 1979. Design of Concrete Structures. Ninth Edition, McGraw-Hill Book Company, New York, NY.

DESIGN OF CFRP-STRENGTHENED CONCRETE MEMBERS

Example 1. Change of Use for a One-Way Joist Slab Floor System.

Building is a 3 story pan joist construction. In the example, standard-pan joists need to be strengthened due to the increase of live load.

Building floor plan:

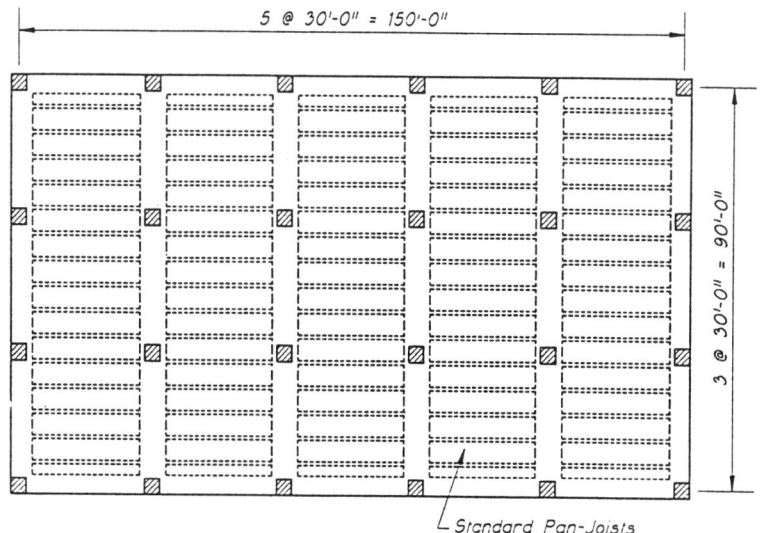

The existing joist section:

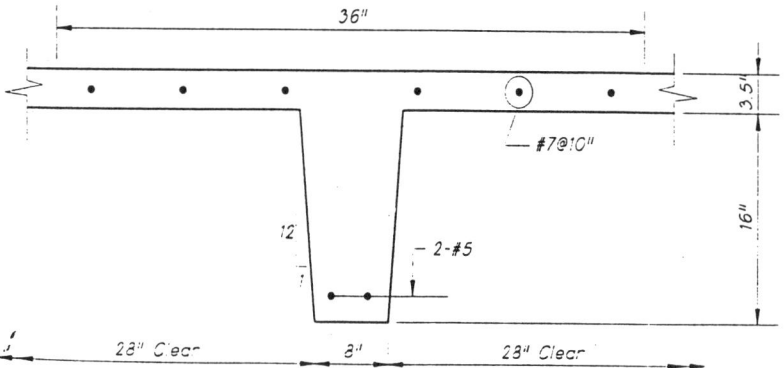

Loads:

LL = 130 psf (increased from 60 psf)
DL = 130 psf (assume 100 psf joists + 20 psf partitions + 10 psf ceiling and misc.)

Calculate moments and shear forces for the joist system using coefficients (simplified design):

Critical span: First interior
Span: 27.5 ft
Width of interior beams: 3 ft
Dead Load: $w_D = (0.13) \times (3) = .39$ klf
Live Load: $w_L = (0.13) \times (3) = .39$ klf
Factored Load: $w_u = 1.4 (0.39) + 1.7 (.39) = 1.21$ klf
Shear Due to DL (exterior): $V_D = w_D l_n/2 = (0.39 \times 27.5)/2 = 5.4$ k
Shear Due to LL (exterior): $V_L = w_L l_n/2 = (0.39 \times 27.5)/2 = 5.4$ k
Shear Due to DL (interior): $V_D = 1.15 w_D l_n/2 = 1.15 \times (0.39 \times 27.5)/2 = 6.2$ k
Shear Due to LL (interior): $V_L = 1.15 w_L l_n/2 = 1.15 \times (0.39 \times 27.5)/2 = 6.2$ k
Pos. DL Moment: $M_D = w_D l_n^2/14 = (0.39 \times 27.5^2)/14 = 21.1$ ft-k
Pos. LL Moment: $M_L = w_L l_n^2/14 = (0.39 \times 27.5^2)/14 = 21.1$ ft-k
Neg. DL Moment (exterior): $M_D = w_D l_n^2/24 = (0.39 \times 27.5^2)/24 = 12.3$ ft-k
Neg. LL Moment (exterior): $M_L = w_L l_n^2/24 = (0.39 \times 27.5^2)/24 = 12.3$ ft-k
Neg. DL Moment (interior): $M_D = w_D l_n^2/10 = (0.39 \times 27.5^2)/10 = 29.5$ ft-k
Neg. LL Moment (interior): $M_L = w_L l_n^2/10 = (0.39 \times 27.5^2)/10 = 29.5$ ft-k
Factored Neg. Moment (exterior): $1.4 M_d + 1.7 M_L = 1.4 (12.3) + 1.7 (12.3) = 38.13$ ft-k
Factored Neg. Moment (interior): $1.4 M_d + 1.7 M_L = 1.4 (29.5) + 1.7 (29.5) = 91.45$ ft-k
Factored Pos. Moment: $1.4 M_d + 1.7 M_L = 1.4 (21.1) + 1.7 (21.1) = 65.4$ ft-k
$1.2 M_d + 1.2 M_L = 1.2 (21.1) + 1.2 (21.1) = 50.6$ ft-k
Factored Shear (interior): $1.4 V_D + 1.7 V_L = 1.4 (6.2) + 1.7 (6.2) = 19.2$ k
Factored Shear (exterior): $1.4 V_D + 1.7 V_L = 1.4 (5.4) + 1.7 (5.4) = 16.7$ k
Factored Shear at d from face: $V_u = 19.2 - 1.21(18.25/12) = 17.3$ k

Using the simplified design method, calculate moment capacity of the unstrengthened section:

Check critical negative moment capacity:

#7 @ 10 in. ($A_s = 0.5$ in²/ft,=1.5 in²):
$a = A_s f_y/(.85 f'_c b) = (1.5)(60)/[(0.85)(4)(8)] = 3.31$ in.
$M_n = A_s f_y/(d - a/2) = (1.5)(60)(18.25-3.31/2) = 1493.6$ in-K = 124.5 ft-k
$\phi M_n = 0.9(124.5) = 112$ ft-k, which is greater than the demand (91.45 ft-k).

Check positive moment capacity:

2 #6 bar: $A_s = 0.62$ in²
$a = A_s f_y/(.85 f'_c b) = (0.62)(60)/[(0.85)(4)(36)] = 0.304$ in.
$M_n = A_s f_y/(d - a/2) = (0.62)(60)(18.25-0.304/2) = 673.2$ in-K = 56.1 ft-k
$\phi M_n = 0.9(56.1) = 50.5$ ft-k, which equals $1.2 M_d + 1.2 M_L = 50.6$ ft-k

The existing section has insufficient positive-moment capacity to withstand the increased live load conditions. The section qualifies for strengthening, since its moment capacity equals 1.2 ($M_D + M_L$), and shear capacity ($\phi Vn = 1.1 \times 0.85 \times 2 \times SQRT(f'_c) \times d \times b = 17.3$ k) equals V_u @ d (17.3 k). The unstrengthened section's flexural capacity is also directly calculated within the design spreadsheet when moments and shear due to dead and live loads are input, as shown on the last page of the example. The shear capacity of the unstrengthened section needs to be calculated manually.

Find the anchorage starting point

The last required hand-calculation, before proceeding with the application of the design spreadsheet, is determination of the anchorage starting point. The factored-load moment diagram is shown in the figure below.

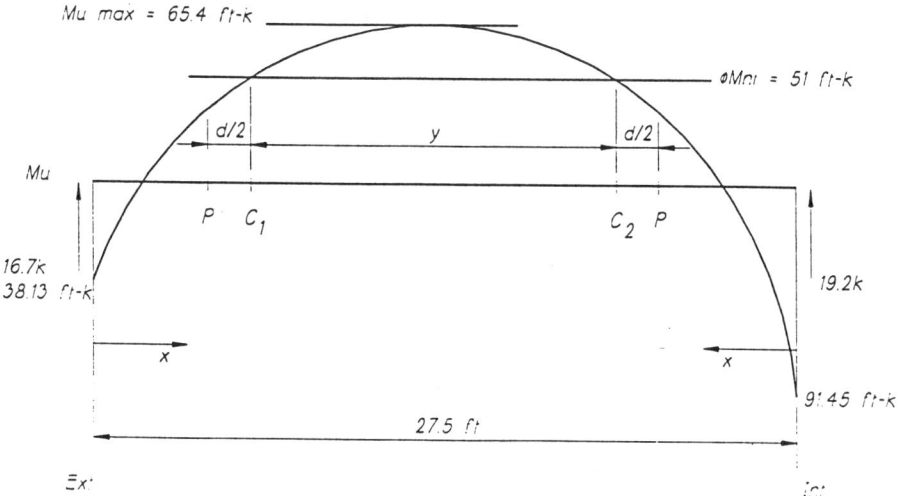

The existing section moment capacity is 51 ft-k, as shown in the figure. The distance from the face of the supports to the intersection points on the moment diagram (x) is calculated as follows:

For exterior side:

Take moments about C_1:

(1.21) $x^2/2$ + 38.13 - 16.7 (x) = -51   and

x = 7.22 ft.

For interior side:

Take moments about C_2:

$(1.21) x^2/2 + 91.45 - 19.2 (x) = -51$ and

$x = 11.81$ ft,

We use x = 7.22 ft*, since it is more critical, and the available anchorage length is then:

$L_{AVAILABLE} = x - d/2 = 6.46$ ft.

*This value is used as the available anchorage-length input on the design spreadsheet.

Design Spreadsheet:

In the left part of the 'User-Input' section of the spreadsheet, all load, span and existing section property information is entered. In addition, this section is assumed "loaded" at the time of strengthening, so by choosing 'Y' in the 'Include initial conditions?' box, the spreadsheet will include section strains due to dead loads in the calculations (during strengthening, only dead loads are expected to be imposed on the section).

In this example, CarboDur S is considered for strengthening, therefore appropriate CFRP material properties are entered in the 'CFRP Properties" section of the design spreadsheet. The last necessary input is the size and number of strips of CFRP. When this part of the design spreadsheet is varied, a number of possible strengthening solutions can be obtained. However, the smaller the amount of reinforcement necessary to develop all design requirements (at any level of ultimate laminate stress), the more economical the solution is. If one strip of 1.4 mm x 60 mm (0.055 in x 2.36 in) CarboDur S is used (as shown in the figure), all design requirements are satisfied at 20 to 40 % of the ultimate stress in the laminate, and the optimal strengthening solution is found.

Example 2. **Strengthening of Foundation Wall to Accommodate a Door Opening.**

In this example a ten-inch foundation wall is cut to facilitate a 10-foot wide door opening leading to the sub grade garage level. The remaining header above the door needs to be strengthened to withstand dead loads from the wall and slab above, as well as live loads from the dwelling units above the garage. The cross section of the header above the door is shown in the following figure:

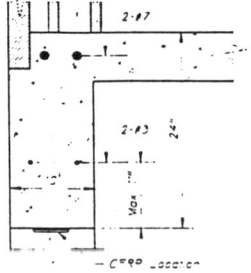

The steel reinforcement shown is the worst possible case, since the horizontal reinforcement in the wall is #3 bars at 12 in. o.c. in each face.

Loads:

			Width or height	
LL	dwelling units	75 psf:	8.75	0.66 klf
DL	7.5 in. concrete slab (7.5/12)(150) =	94 psf	8.75	0.82 klf
	MEC	10 psf	8.75	0.09 klf
	Wall above ground level:	80 psf	9.0	0.72 klf
	Beam (10/12)(50) =	125 psf	2.0	0.25 klf
	Total DL			1.88 klf

Calculate moments and shear forces in the header (fixed-end conditions):

Factored load: $w_u = 1.4\,w_D + 1.7\,w_L = 1.4\,(1.88) + 1.7\,(.66) = 3.75$ klf
Factored negative moment: $M_u = w_u l^2/12 = 3.75\,(10)^2/12 = 31.3$ ft-k
Positive DL moment: $M_D = w_D l^2/24 = 1.88\,(10)^2/24 = 7.83$ ft-k
Positive LL moment: $M_L = w_L l^2/24 = 0.66\,(10)^2/24 = 2.75$ ft-k
Factored positive moment: $M_u = 1.4\,M_D + 1.7\,M_L = 1.4\,(7.83) + 1.7\,(2.75) = 15.6$ ft-k
$1.2\,(M_D + M_L) = 1.2\,(\,7.83 + 2.75\,) = 12.7$ ft-k
Shear Due to DL: $V_D = w_D l/2 = 1.88\,(10)/2 = 9.4$ k
Shear Due to LL: $V_L = w_L l/2 = 0.66\,(10)/2 = 3.3$ ft-k
Factored shear: $V_u = 1.4\,V_D + 1.7\,V_L = 1.4\,(9.4) + 1.7\,(3.3) = 18.8$ ft-k
Factored shear at d: $V_u = 18.8 - 3.75\,(22/12) = 11.9$ k

Calculate moment capacity of the unstrengthened section:

Negative moment capacity:

$A_s = 1.2$ in^2 (2 #7 bars)
$a = A_s f_y/(.85 f'_c b) = (1.2)(60)/[(0.85)(4)(10)] = 2.12$ in.
$M_n = A_s f_y/(d - a/2) = (1.2)(60)(22-2.12/2) = 1507$ in-K $= 126$ ft-k
$\phi M_n = 0.9(126) = 113$ ft-k, which is greater than the demand.

Positive moment capacity:

$A_s = 0.22$ in^2 (2 #3 bars)
$a = A_s f_y/(.85 f'_c b) = (0.22)(60)/[(0.85)(4)(10)] = 0.39$ in.
$M_u = A_s f_y/(d - a/2) = (0.22)(60)(13-0.39/2) = 169$ in-K $= 14.1$ ft-k
$\phi M_n = 0.9(14.1) = 12.7$ ft-k

The existing section has insufficient positive-moment capacity to withstand the existing load conditions (12.7 < 15.6). The section qualifies for strengthening, since its moment capacity equals 1.2 ($M_D + M_L$), and shear capacity ($\phi V_n = 0.85 \times 2 \times \mathrm{SQRT}(f'_c) \times d \times b = 23.7$ k) exceeds V_u@d =11.9 k. The unstrengthened section's flexural capacity is also directly calculated within the design spreadsheet, as shown on the last page of the example. The shear capacity of the unstrengthened section needs to be calculated manually.

Find the anchorage starting point:

The factored-load moment diagram is shown in the figure below:

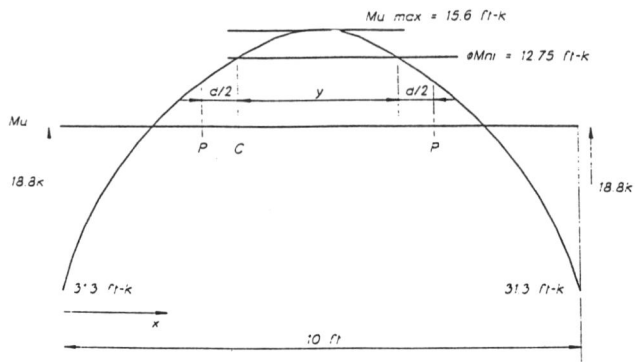

The distance from the face of the supports to the intersection points on the moment diagram (x) is calculated as follows:

Take moments about C:

(3.75) $x^2/2$ + 31.3 - 18.8 (x) = -12.7   and

x = 3.7 ft *

*This value is used as the available anchorage-span input on the design spreadsheet.

Therefore, the available anchorage length is :

$L_{AVAILABLE}$ = x - d/2 = 3.7 - 22/24 = 2.78 ft.

If this calculated distance was larger so that when development length was added it wouldn't fit the underside of the door header, the reinforcement would have been placed on one or both vertical sides of the header above the door.

In this example, CarboDur S is considered for strengthening, therefore appropriate CFRP material properties are entered in the 'CFRP Properties" section of the design spreadsheet. The last necessary input is the size and number of strips of CFRP. When this part of the design spreadsheet is varied, a number of possible strengthening solutions can be obtained. However, the smaller the amount of reinforcement necessary to develop all design requirements (at any level of ultimate laminate stress), the more economical the solution is. In this particular example, the strengthening demand was fairly small. Therefore, if one strip of 1.2 mm x 50 mm (0.047 in x 1.97 in) CarboDur S is used (as shown in the figure), all design requirements are satisfied at 20 % to 40 % of the ultimate stress in the laminate, and the optimal strengthening solution is found.

Upgrade of Naval Wharf Bravo 25, Pearl Harbor, Hawaii

S. LANSBURG

OBJECTIVE

The Navy's objective was to make concrete repairs, add cathodic protection, and bond external CFRP reinforcing to two segments of Bravo 25 at the Naval Station Pearl Harbor. This upgrade will provide reinforced "platforms" to support crane outrigger loads up to 110 kip (980 kN) and a uniform load up to 750 psf (35 kPa).

This project is part of an ongoing Naval Facilities Engineering Service Center (NFESC) Repair & Upgrade Program to demonstrate advanced technologies for post strengthening existing Navy piers. The methodology incorporates the placement of a few ounces/ft^2 of high strength material (CFRP) in strategic locations versus massing hundreds of pounds of additional concrete & steel. The simplicity of the upgrade allows only minimal interruptions in pier operations as opposed to the pier closing for up to a year for a traditional alteration. This upgrade defines and demonstrates the application of post strengthening using fiber-reinforced plastic systems to meet demands due to operational changes for which the original pier design did not include.

Existing Conditions: Original construction is cast in place reinforced concrete deck and superstructure supported by precast concrete piles. The Bravo wharves are over 50 years old. Bravo wharves were originally designed to support 50 ton (445 kN), rail mounted portal cranes, train cars, as well as 900 lbs per sq. ft. (42 kPa).

Track mounted cranes have been replaced by truck mounted, mobile cranes. Mobile crane current limitations currently placed on Bravo are very restrictive. Other areas are limited to trucks and forklifts. Maximum uniform live load is limited to 490 psf (kPa).

ACE Restoration & Waterproofing, Inc., 740 E. Walnut Ave., Fullerton CA 92831, USA

SCOPE OF WORK

Work included concrete repair (epoxy injection, crack sealing, spall repair), an impressed current cathodic protection installation, and strength upgrades on selected spans.

Cathodic Protection System: Cathodic Protection (CP) may be the only practical means of mitigating corrosion of concrete reinforcing steel once it has started. Alteration of natural electrochemical processes that result in corrosion can only be achieved by driving reactions in favorable directions on the concrete surface. This is the methodology of CP.

An impressed current CP system was installed to mitigate future corrosion. This work involved slotting the underside of the deck, and installing a titanium ribbon anode into the slots without the ribbon making contact with the reinforcing steel. Rebar cover is sufficient to achieve an adequate current distribution for the embedded CP system.

The required voltage is less then 30 volts. The anode configurations are to provide uniform current distribution over the surface of each installation.

Lead wires were connected to three reinforcing steel bars in each CP Grid. Continuity test and connection resistance tests were conducted in all areas.

CFRP Reinforcement: his work involved positive moment, longitudinal reinforcement to the bottom of the 13 ½ (34 cm) and negative moment, transverse reinforcement to the top of the deck. .

The Navy's selection criteria for external reinforcing are stiffness, strength, flexibility, and reliability. High strength carbon/epoxy composites were selected because it is non-corrosive, alkaline and chemical resistant, with high strength and stiffness to control deflections and cracking. Of all fibers, high strength carbon possesses some of the best resistance to the extremes of the maritime industrial environment.

Required upgrade reinforcing areas were based on an equivalent working stress of 200 ksi on the carbon fiber at a strain of 0.7 percent, which recognizes strength losses with time and repeated load application. High strength carbon fiber area of 0.033 in^2/in-width was required for the bottom surface, while 0.018 in^2/in-width was required for the areas in the top surface. For a composite with 65 percent carbon fiber this translates to a laminate area of 0.051 and 0.028 in^2/in width respectively.

Negative Moment Embedded Reinforcement: Embedding pultruded carbon/epoxy composite rods as negative reinforcing is particularly attractive for the top of the deck where exposed external reinforcement would be subject to mechanical and environmental damage.

These pultruded laminates contain 60 to 65 percent high strength carbon fibers in an epoxy matrix. The rods will be encapsulated in a two-part amine epoxy embedded in slots cut in the concrete surface. The

3/8-inch rods were embedded into 1 inch (25mm) deep and ½ inch (12mm) wide, spaced 4 inches (10 cm) on center.

Positive Moment External Bonded Wet Lay Up Composite Laminate: A uniaxial carbon fiber sheets in a epoxy resin matrix (saturate) having a tensile strength of 3.3 kips/inch-width (5.8 kN/cm-width) and a fiber weight of 0.06lb/ft^2 (300 g/m^2). 5 layers were installed to achieve the carbon fiber area of 0.033 in$^\square$/in-width.

CONCLUSIONS:

This project was unique in the fact that a Composite Upgrade was installed over a Impressed Current Cathodic Protection System. The busy wharf was kept in operation during the construction, which was of importance to the Navy.

Extensive testing before, during & after the upgrade was performed by the Navy, including load testing. A report is forthcoming from the Navy, which will be public record.

Structural Upgrade and Repair of Wood Members Using Cross-Ply Carbon/Epoxy

A. MOSALLAM, J. KREINER and T. LANCEY

ABSTRACT

Results of a pilot research study on the use of polymer composites in repair and rehabilitation of structural wood members are presented. In this program, a total of eight full-scale tests were conducted to investigate the behavior of wooden members repaired and retrofitted with external composite laminates. A total of eight 8" X "8 X 10' *(203 mm X 203 mm X 3.0 m)* Douglas Fir Larch # 1 wooden members with and without polymer composites overlays were tested to failure. The control specimen tests were conducted to determine the ultimate moment capacity of the unreinforced wood. A control damaged (pre-cracked) specimen was subjected to a similar loading regime to compare its performance to a similar beam with a carbon/epoxy composite repair system. For repair application, a damaged beam was repaired by applying two cross-play carbon/epoxy laminates covering both the tension side and two-thirds of the beam depth and was loaded to failure. Load/deflection and stress/strain curves were developed and modes of failure were identified. For the rehabilitation application, an "undamaged" wood beam specimen was strengthen externally with two laminates of cross-ply carbon/epoxy composites, and was loaded to failure. The ultimate capacity of the unreinforced wood member was 7.1 kips (31.6 kN) with a deflection of 1.5" (38 mm) at failure. The load carrying capacity of the retrofitted specimen has increased up to 238.27% and the initial stiffness was increased by about 700% as compared to the control specimen. The associated failure mode was sudden and was initiated by a cohesive failure of the wood at the specimen side exactly under the left loading point. Test results also indicated that, adding external laminates of polymer composites to the damaged "pre-cracked" specimen resulted in an appreciable increase in the loading carrying capacity up to 180%, with an average initial stiffness upgrade to about 110%.

INTRODUCTION

Wood is one of the widely used construction materials, especially in the housing industry in the USA. In addition, numerous historical bridges were built in the last two

---

Division of Engineering, California State University-Fullerton, Fullerton, CA 92834, USA

centuries using wood products. Increasing the structural efficiency of wood structures will not only contribute in advancing the engineered wood science and decreasing the initial cost, but also will contribute in tremendous saving of our environment by decreasing the number of trees to be cut every day to satisfy the increasing demand of the construction industry worldwide.

Recently, polymer composites have been considered by the majority of the construction community as a valuable material for repair and rehabilitation of infrastructure systems [1]. As a result of the successful experience with these materials in other structural applications, several research studies were lunched to explore the possibility of using these materials to produce high-performance engineered wood products. Some of the pioneering work was reported as early as 1965 by Biblis [2] and Theakston [3]. Biblis [2] evaluated the performance of glass fiber composites used as face sheet materials of wood-core sandwich members. Similar work was also reported by Boehme and Schulz in 1974 [4]. Theakston [3] work is considered one of the pioneering work on studying wrapped wood members. Similar concept proposed by Theakston [2] was investigated recently by Arnold and Fyfe [5] where E-glass/Epoxy laminates were used to restore damaged wooden utility poles. Another application was commercially introduced by Tingley [6] in fabricating a new hybrid glued-laminated wood by introducing thin laminates of E-glass composites between the laminated wood. Over 1,000 full-scale glued-laminated wood beams using reinforced polymers have been manufactured and tested. Recently, several papers were published discussing the critical issues related to this application. For example, in 1999, Davalos et al. [7] discussed both conventional and fracture mechanics tests for evaluation of wood-FRP interfaces. In 1991, Triantafillou and Deskovic [8] presented the results of a study on the use of carbon/epoxy composites for upgrading the structural capacity of wood members.

MOTIVATION AND OBJECTIVE

Limited experimental full-scale information is available on structural behavior of full-scale wood members reinforced with polymer composites. In order to introduce this application to the construction industry, more experimental results in both the short- and long-term behavior of composite repaired wood are essential. In this pilot project, a total of eight full-scale tests were conducted to investigate the behavior of wooden members repaired and retrofitted with external composite laminates.

EXPERIMENTAL PROGRAM

General: The specimens tested in this program were 8" X "8 X 10' (203.2 mm X 203.2 mm X 3.048 meters) Dug Fur Larch No. 1 wooden members. According to the NDS, the allowable bending stress for these members is 700 psi (4.83 MPa). All specimens were tested to failure under four-point load regime (see Figure (1)). Initially, two control specimens were tested under quasi-static loading conditions. The other two beam tests were conducted on repaired and retrofitted wood specimens with same materials and dimensions. All the beams were instrumented with electrical strain gages at several locations along the beam span and depth at both wood and composite surfaces. The loading rate was 1.0 kips/minutes (4.448 kN/minute), and the frequency of strain, deflection, and load information was at a rate of 10 readings per second.

Unreinforced Control Test: The first unreinforced control specimen was tested to failure to determine the wood beam ultimate moment capacity, flexural stiffness, and to identify the failure mode of unreinforced wood member. As the load increased, longitudinal cracks with an average length of ¾" (19 mm) were observed. The failure mode of the control specimen was sudden. The crack failure line initiated at a knot located at the bottom side of the member below the loading points as shown in Figure (2). The ultimate mode of failure was a combination of interlaminar and tensile failure of wood between the load points. The ultimate load was 7.05 kips (31.36 kN) with a maximum deflection of 1.5" (38 mm). This specimen exhibited a linear behavior up to a load of 3.0 kips (13.34 kN) after which non-linear behavior was observed as shown in Figure (7). The initial stiffness of the control specimen was 1.16 Msi (8.0 GPa) and the ultimate strength of 3.16 ksi (21.79 MPa). The strain at failure of the control specimen measured at the extreme fibers of the mid-span tension side was 0.44%.

Figure (1): Four-point Bending Setup for Wood Members

Unreinforced "Pre-cracked" Test: In order to demonstrate the effectiveness of the composite system in restoring damage wood members, a second control specimen "pre-cracked" specimen was subjected to a similar loading regime. The damage was simulated by introducing 1/8" (3.18 mm) thick-blade saw cuts spaced at 5" (127 mm) at the constant moment area (between the two concentrated loads) with a constant cut depth of 2.75" (69.85 mm) as shown in Figure (3). Strain gages were bonded to sides and the top surface of the wooden member at several locations (*Unlike other specimens, no strain gages were applied to the mid-span bottom surface of this specimen. For this reason, no flexural tensile stress data is shown in Figure (7)*). Same reading schedule and loading rate were used similar to those for the control specimen. As the load was applied, large deflection was observed. This was expected due to the fact that by slotting the bottom area under the loading points (maximum constant moment zone), the effective resisting depth was about 65% of the corresponding depth in the control test. The deflection continued to increase and longitudinal cracks propagated in a higher rate. The ultimate failure load was 8.40 kips with a corresponding deflection of 4.65" (118 mm).

Figure (4) shows the ultimate failure of the pre-cracked specimen. The load/deflection relation is presented in Figure (5). One interesting observation is that the ultimate load of the pre-cracked specimen was higher that the ultimate load of the control specimen. This can be attributed to the variation of the wood materials, locations and number of knots (*the failure of the control specimen started at a knot location as shown Figure*).

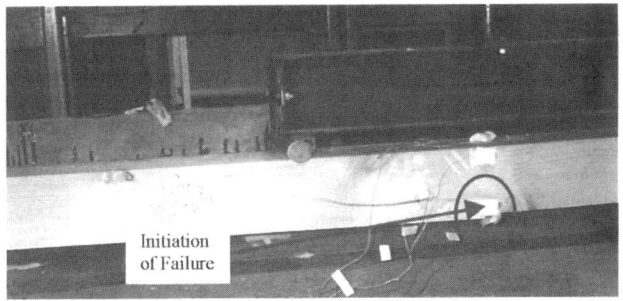

Figure (2): Failure was initiated at a Knot Located at the Bottom of the Loading Zone

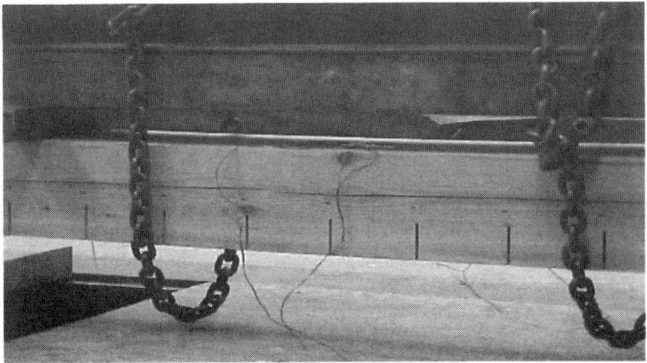

Figure (3): 1/8" (3.18 mm) Vertical Slots were introduced at 5" (127 mm) Spacing under the Loading Points

Figure (4): Ultimate Failure of the Pre-Cracked Specimen

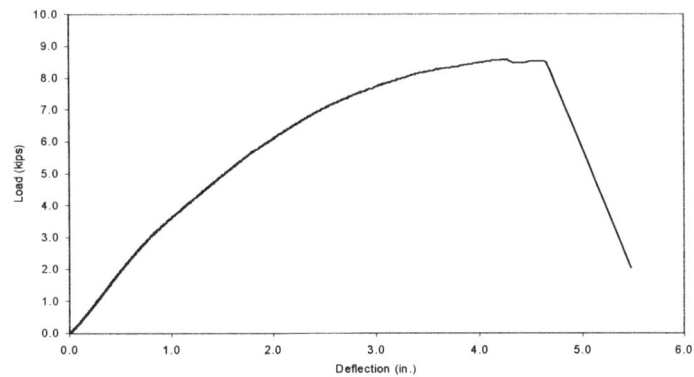

Figure (5): Typical P/δ Curve for Control Specimens

3.4 Repaired, Pre-cracked Beam Test: This test was conducted on another "pre-cracked" specimen. The specimen was instrumented was strain gages at different critical locations. First, the cracks were cleaned and filled with epoxy (to about have crack depth), and was allowed to cure. The bottom and sides surfaces were prepared, and cleaned. A thin film of FiberBond® low-viscosity primer with applied. The composite lamination used for this test was composed of FiberBond® bi-directional woven carbon fabrics and high-strength/high-toughness epoxy system $(0°/90°)_{2s}$. The carbon/epoxy laminates were applied to the tension side of the wood member and the sides were extended to cover 75% of the pre-cracked specimen depth (from the bottom). The reason of using this design was based on the observations of crack propagation and ultimate failure mode observed in the control test. The longitudinal fibers were designed to carry the flexural tensile stresses, while the vertical fibers were designed to resist the interlaminar shear stresses. Strain gages were bonded to wood and composite surfaces at different critical locations and data were collected automatically using a computerized data acquisition system as for the other two specimens. The behavior of this specimen was linear up to about 5 kips (22.25 kN), after which a slight non-linearity was observed (see Figure (7)). It is important to point out that this specimen did not fail, since the test was halted due to the hydraulic actuator stroke limitation. The maximum load recorded in this test was 18.57 kips (82.5 kN) with a corresponding mid-span deflection of 3.04" (77.22 mm). The longitudinal axial tensile strain of the fibers at the maximum load (not failure load) was 0.17%. The corresponding maximum tensile stress at the mid-span was 6.92 ksi (47.7 MPa). As compared to the strength of the control specimen (σ_u= 3.16 ksi (21.79 MPa), an increase in the ultimate strength of about 117% was achieved by adding the two layers of FiberBond® cross-ply laminates. It is also important to note that stress and strain values for this specimen are not the ultimate values and higher values are expected.

Retrofitted Beam Test: Same lamination schedule was adopted in strengthening the "undamaged" wood specimens. Strain gages were bonded at both the wood and composite surfaces. The specimen was tested under the same load setup as for the other three specimens. This specimen exhibited stiffer behavior as compared to the other three specimens, and had a bi-linear behavior. The first linear part was maintained up to a load level of 13 kips (57.82 kN), while the second near linear stiffness was observed until a load of 23 kips (102.3 kN), after which the yield was achieved and maintained until failure occurred (see Figure (7)). The associated failure mode was less sudden in nature as compared to the one observed for the control specimen. The ultimate failure mode was initiated by a cohesive failure of the wood at the specimen side, exactly under the left point load, followed by a local buckling of the thin-walled composite laminate at the same location (see Figure (6)). The stress/ strain curves for pre-cracked, repaired, and the retrofitted wood members are presented in Figure (8). Based on the test results, the gain in the strength of the retrofitted members as compared to the control members is in the order of 238.27%. The initial stiffness of this specimen was about 771% of the initial stiffness of the control "unreinforced" wood specimen. The ultimate stress and strain at failure were 8.86 ksi, and 0.67%, respectively.

Figure (6): Ultimate Failure Mode of the Retrofitted Specimen

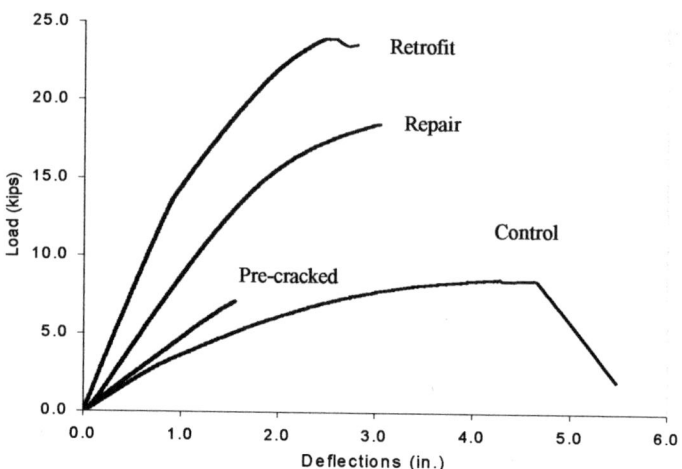

Figure (7): Load/Deflection Curves for All Specimens

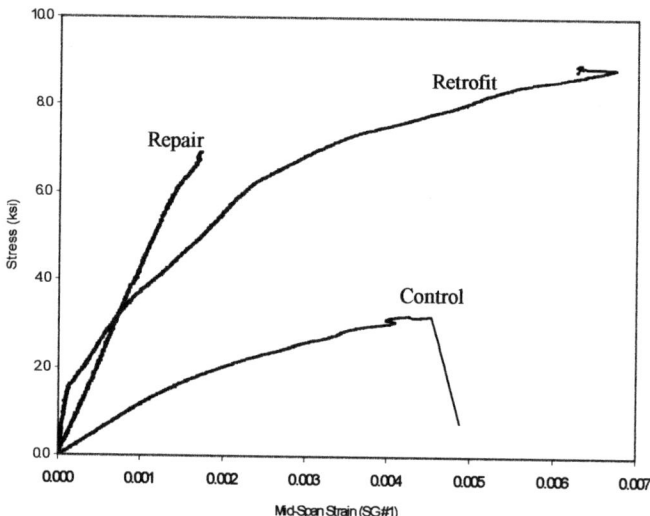

Figure (8): Stress/Strain Curves for Pre-cracked, Repaired, and Retrofitted Specimens

SUMMARY & CONCLUSIONS

Based on the experimental results, it evident that the use of carbon/epoxy composite, manufactured by Edge Structural Composites, has performed exceptionally will in not only restoring the original capacity of the damaged wood members, but also increased the original strength dramatically as described earlier. Table (1) summarizes the experimental test results obtained from this program. A graphical representation of Table (1) is shown in Figure (9).

Table (1): Summary Results of the Wood Experimental Program

SPECIMEN	ULTIMATE LOAD, kips (kN)	ULTIMATE STRESS, psi (MPa)	DEFLECTION @ ULTIMATE, inch (mm)	PERCENTAGE OF STRENGTH INCREASE (%)
Control	8.47 (37.67)	2,432 (16.76)	1.55 (39.37)	--
Pre-Cracked	7.08 (31.49)	2,032 (14.01)	4.65 (118.16)	--
Repaired	18.54* (82.46)	5,331 (36.76)	3.04 (77.21)	180**
Retrofitted	23.95 (106.53)	6,876 (47.41)	2.80 (71.12)	238***

* Specimen did not fail,
** As compared to the ultimate strength of the pre-cracked specimen
*** As compared to the control specimen

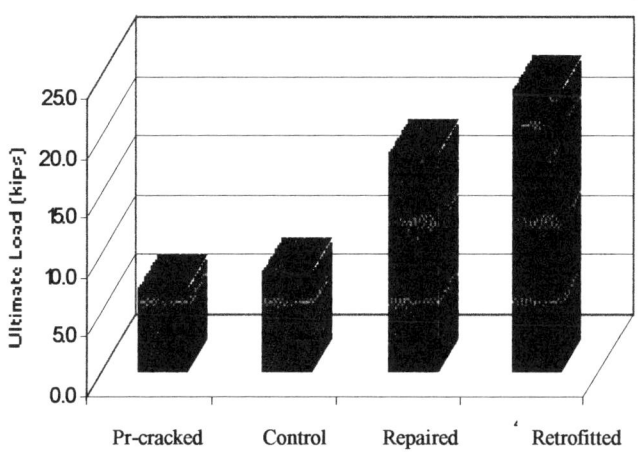

Figure (9): Ultimate Load Comparison

ACKNOWLEDGMENT

The research project described in this paper is a part of the ICBO/Caltrans Structural Evaluation of FiberBond° System under two research contracts with Edge Structural Composites, LLC conducted at the SRRS Center of California State University, Fullerton. The authors would like to acknowledge the great effort of Mr. James Kiech in operating the tests.

REFERENCES

1. Mosallam, A.S., Editor (1998). <u>Advanced Seismic Repair & Rehabilitation Structural Systems</u>, Proceedings, SRRS1, Fullerton, California, November, 201p.
2. Biblis, E. J. (1965). *"Analysis of Wood-Fiberglass Composite Beams within and Beyond the Elastic Region,"* Forest Production Journal, Volume 15, No. 2, pp. 81-88.
3. Theakston, F. H. (1965). *"A feasibility Study for Strengthening Timber Beams with Fiberglass,"* Canadian Agricultural Engineering, January, pp. 17-19.
4. Boehme, C., and Schultz, U. (1974). *"Load Bearing Behavior of a GFRP Sandwich,"* Holz Roh – Werkst, Vol. 32, No. 7, pp. 250-256.
5. Arnold, S. F., and Fyfe, E. R. (1998). *"Repair and Rehabilitation of Wooden Members Using Polymer Composites,"* Advanced Seismic Repair & Rehabilitation Structural Systems, Edt. Ayman Mosallam, Proceedings, SRRS1, Fullerton, California, November, pp. 52-56.
6. Tingley, D. (1999). "*Reinforced Plastic Computability with Wood Composites,*" Proceedings, 5th ASCE Materials Engineering Congress, pp.108-115.
7. Davalos, J., Qiao, P., and Trimble, B. (1999). "*Conventional and Fracture Mechanics Tests for Evaluation of Wood-FRP Interfaces,*" Proceedings, 5th ASCE Materials Engineering Congress, Cincinnati, OH, pp.108-115.
8. Triantafillou, T., and Deskovic, N. (1991). *"Innovative Prestressing With FRP Sheets: Mechanics of Short Term Behavior,"* Journal of Engineering Mechanics, ASCE, 117 (7), pp. 1625-1672.

Settlement Repair of Lightly Reinforced Concrete Block Walls Using CFRP

G. MULLINS, A. HARTLEY, D. ENGEBRETSON and R. SEN

ABSTRACT

This paper presents the results of an experimental study to assess the feasibility of using uni-directional carbon fiber reinforced polymer sheets for repairing lightly reinforced concrete masonry walls that have cracked due to foundation settlement. In the study, two full-sized 8 in. block walls, 8 ft high and 20 ft long were first damaged under simulated settlement loading and then re-tested after being repaired using carbon fiber sheets bonded to one side of the wall. Strength gains of over 50% were recorded. This suggests that carbon fiber reinforced polymers may be suitable for repairing concrete block walls damaged by foundation settlement.

INTRODUCTION

Un-grouted, lightly reinforced concrete block walls are widely used in residential construction in southeastern United States where wind, not earthquake loading, is critical. Until as recently as the early 1990's, walls were vertically reinforced by #5 bars at the corners or every 240 sq. ft. (i.e. every 30 ft, for typical 8 ft high walls) and by a horizontal #5 bar in 8-in. bond beams. This is in stark contrast to masonry construction in seismic regions that are fully grouted with significant reinforcement in both the vertical and horizontal directions.

The vast majority of masonry homes are fortunately not subjected to hurricanes (or even earthquakes) in their lifetime. Instead, they stand a greater chance of being damaged by foundation settlement. Nationwide, foundation subsidence affects an estimated quarter-million homeowners *annually* [1]. Conventional repair of settlement damage can be both time consuming and costly. In contrast, fiber reinforced polymer (FRP) materials offer the prospect of speedy, unobtrusive and inexpensive repairs. This paper describes the results from a pilot study to investigate the feasibility of using FRP material to repair foundation settlement problems in lightly reinforced masonry walls.

Gray Mullins and Rajan Sen, Department of Civil and Environmental Engineering, University of South Florida, Tampa, FL 33620
Alfred Hartley, Carlan Killam Consulting Group Inc., Tampa, FL 33619
Dan Engebretson, Walter P. Moore and Associates, Inc., Tampa, FL 33602

Figure 1. Characteristic Settlement Cracking.

BACKGROUND

Foundations for masonry walls (in the southeast) are commonly reinforced longitudinally by two #5 bars to allow the footing to bridge over soft nonuniform soil. However, improper compaction or the presence of buried compressible material such as tree roots or construction debris can nonetheless lead to excessive settlement. As unreinforced masonry is brittle it has a greater propensity to crack even when the movement is within the allowable limit for the concrete footing.

Characteristic settlement cracks observed in block homes are stair-step mortar cracking or horizontal mortar cracks between the concrete floor slab and the first course of blocks as shown in Fig. 1. The failure is the result of un-grouted masonry's very low tensile strength - allowable values vary between 25-50 psi [2] - compared to modulus of rupture values of over 400 psi for 3000 psi concrete [3].

Typically, repairs involve two steps. *First,* the movement of the foundation is stabilized either through the use of grade beams or a combination of both grade beams and pin piles. *Second*, the wall is transformed to a 'wall-beam' by the introduction of vertical steel reinforcement (Fig. 2) or fully grouting all the cells. The latter option provides a nearly eightfold increase in shear capacity. The cost of repairing an average home with 140 ft length of masonry wall is in the range of $40-45,000 [4].

OBJECTIVES

The primary goal of this study was to investigate the feasibility of using uni-direction fiber-reinforced polymer material to repair walls damaged by foundation settlement. An important element was to examine the performance of as-built walls that conformed to construction standards since the 1960's so that the results of the study would be more widely applicable.

Figure 2. Conventional Repair of Settlement Damage.

SBC PROVISIONS

The governing design code for the southeast is the Standard Building Code (SBC) [5]. As part of this study, a detailed historical review of the relevant masonry and wind provisions of SBC was carried out to identify construction practice since the first edition of the code was published in 1946. The findings may be summarized as follows: (1) no vertical reinforcement - 1940's through 1950's construction; (2) vertical reinforcement only at the corners - 1960's through early 1990's; and (3) intermediate reinforcement required - 1993 to present.

Intermediate reinforcement was actually required starting with the 1969 edition where masonry could "... not exceed 240 square feet without approved vertical and horizontal support..." for walls which were 8 in. thick. However, this revision specifically excluded "one and two family residences". By the 1988 SBC Edition, this exemption was removed but the fact that most single-story masonry walls were only eight feet high meant that intermediate reinforcement was still only required every 30 feet. At best, a continuous masonry wall may have required one intermediate rebar; and this revision had little impact on single-story residential walls. The deemed to comply "Standard for Hurricane Resistant Residential Construction (SSTD 10)," first published in 1993 [6], further reduced the spacing for vertical reinforcement. For example, SSTD 10-93 required that a 44 ft x 8 ft x 8 in masonry wall have intermediate reinforcement at a maximum spacing of four feet and a rebar aside every opening to withstand 110 mph wind velocities.

TEST PROGRAM

A careful survey was made of subsidence damage to masonry block homes in the Tampa Bay area prior to the design of the test specimens. Based on this inspection and interviews with consulting engineers, the representative wall length in testing was determined to be 20 ft. The wall height was taken as 8 ft which is typical for residential construction.

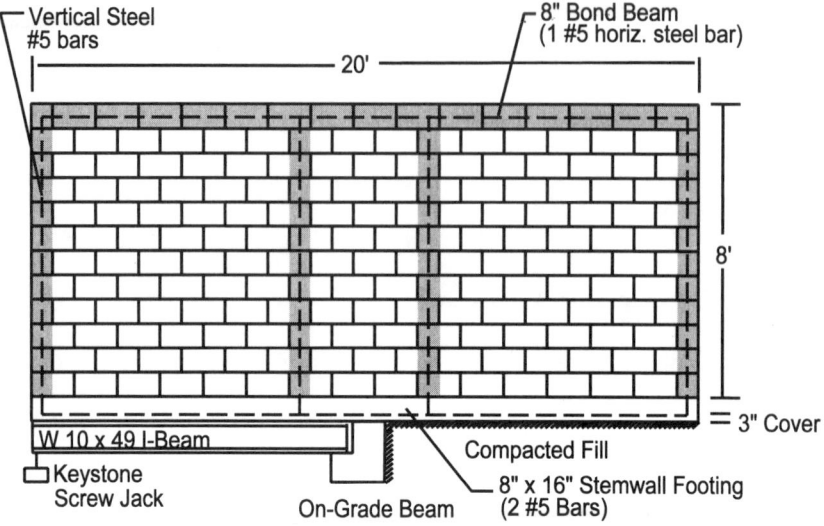

NOTE: Intermediate Vertical Reinforcement Omitted in Wall 3

Figure 3. Reinforcement Details - Wall 2.

A total of four walls were constructed by experienced local masonry contractors to represent the prevailing building standards from the 1960's to the present. Walls 2 and 3 represent construction practice prior to the 1990's that provided vertical reinforcement only at the corners and optionally at intermediate locations; Walls 1 and 4 represent current practice with vertical reinforcement at 4 ft spacing. Wall 2 had intermediate reinforcement spaced at 8 ft from the ends (see Fig. 3) while Wall 3 had no intermediate reinforcement (ends only).

Past and current building practices were represented by two different methods of forming the bond beam. In the older style of construction, a 6 in. wide felt strip was applied to the last course of the masonry unit just prior to the bond beam course. The felt strip covers the open cells of the masonry units and prevents grout from falling through while the bond beam is poured. The felt cap, however, reduces the face shell bedding for the typical 8 in. block from 1.25 in. to only 13/16 in. width for each shell. Current practice is to cover each cell with a thin metal pan. Walls 1 and 4 were constructed with metal fill caps while Walls 2 and 3 were built with a felt cap.

The four walls were built such that they shared a common support wall that was perpendicular to each wall. All of the walls were built on top of a 8 in x 16 in concrete footing reinforced with two #5 bars. Dowels with 90° hooks were provided at the location of the vertical steel to connect wall to the foundation. Each footing was cantilevered approximately 10 ft. 6 in. past a concrete grade beam. This beam served mainly to retain the soil and support the steel beam - the footing did not bear upon the grade beam.

The interface between the test walls and the support wall was a butt rather than an interlocking joint that is typical of actual construction. An interlocking joint would have provided greater support to the test walls at the interface and prevented rotation of the wall panel as a whole. However, the bond beam was poured continuously such that the common grout core shared by all of the walls provided some degree of support. The grouting of the reinforced cells and pouring of the bond beam was performed in one operation.

Test Setup

Differential settlement cracking has been observed to occur only over a portion of a masonry wall. This suggests that part of the wall is continuously supported while the rest is free to subside. This condition was simulated in the test setup by constructing about half the length of the wall on compacted fill. The remainder was supported on a W10x49 steel beam that cantilevered off a grade beam. This steel beam was supported at its free end on a keystone screw jack (see Fig. 3). Plastic sheeting was used to separate the footing from the steel beam and prevent composite action.

Three complete turns of the hex-nut in the screw jack provided 3/8 in. settlement which was sufficient to allow a 10 ft, 6 in. length of the wall to act as a cantilever under its own weight. Additional settlement was induced by applying loading at the top of the wall using a 50-ton ENERPAC hydraulic jack. A steel reaction frame made of W12 x 22 steel beams provided the reaction to the jack. The frame was held down by chains connected to "Chance"anchors bored into the soil. Each was drilled to a depth of 12 ft to 15 ft and was located on both sides of each wall. The reaction frame geometry and the location of the ground anchors permitted loads of up to 15 tons to be applied to the cantilevered end of each wall (see Fig. 4).

Instrumentation

The test specimens were instrumented using load cells, LVDT's, and strain gages. Two load cells were used - one at the free end to measure the self weight when the screw jack was released and the second to measure the magnitude of the simulated roof load. LVDT's measured deflections in both the vertical and horizontal directions; strain gages were used to record strains in both the wall and the footing. A battery powered MEGADAC 3108 DC data acquisition system was used for recording data.

Test Procedure

Two series of tests were carried out. In the first series, the settlement loads consisting of the self weight of the cantilevered wall and simulated roof loads were applied that led to the development of stair-step cracks. Subsequently, the loads were removed and the hex nuts of the screw jack returned to their original position. The

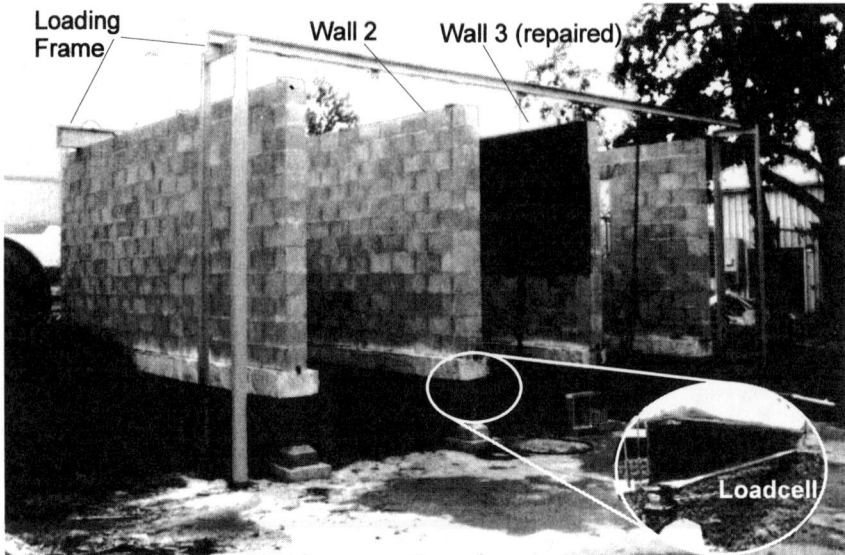

Figure 4. View of Wall Set-Up.

wall was then repaired using carbon fiber reinforced polymer sheeting. As the surface of the walls was smooth, no special surface preparation was deemed necessary and manufacturer's recommendations were generally followed. The ends of the CFRP where secured to the wall with 3 in. steel plates using Tapcon anchors.

Since the uni-directional carbon fiber sheeting (FTS-C1-120) was only intended to increase bending resistance, it was oriented parallel to the length of the wall. Three sheets, each 20 in. wide, were attached *only to one face* as shown in Fig. 4. The intent was to provide additional bending and shear capacity in the zone of the wall that was expected to be in tension. Strain gages were mounted on the CFRP surface and the strengthened wall tested in the same manner as the plain wall. As the specimens were constructed outdoors, all tests were carried out in the pre-dawn hours to ensure constant thermal conditions during the testing.

RESULTS

Two of the four walls constructed have been tested. These were the two weakest walls (#3 with vertical reinforcement at the ends) and (#2 with vertical reinforcement 8 ft from the ends). The results obtained included crack patterns, load deflection and strain variation with depth. Because of space limitations only results relating crack pattern and load deflection are presented.

The crack pattern from the two walls are shown in Figs. 5 and 6. Note the stair step cracking in the supported section of the wall caused by the rotation of the cantilevered

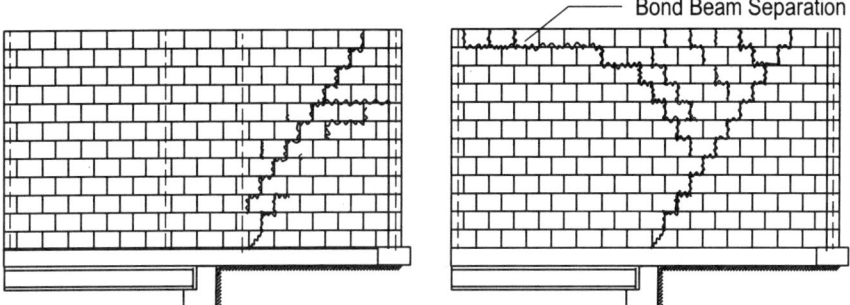

Figure 5. Crack Pattern, Wall 2.   Figure 6. Crack Pattern, Wall 3.

section. The cracking was more pronounced in the weaker wall (#3) as there was no intermediate vertical steel. In contrast, cracking in the stronger wall was confined to the supported region. This is not surprising given that the cantilevered portion of the structure was somewhat longer.

The effect of repair on the capacity of the walls is shown in Figs 7 and 8. Each figure displays the response of plain and strengthened walls. As the weaker wall cracked at a proportionately smaller load, its strength gain was greater (72% vs 57%). Inspection of Figs. 7 and 8 also show improvement in the stiffness of the repaired wall. This was evident in the load vs strain plot. The highest strains were in the bond beam and in the footing. Strains in the remainder of the wall were relatively modest [7].

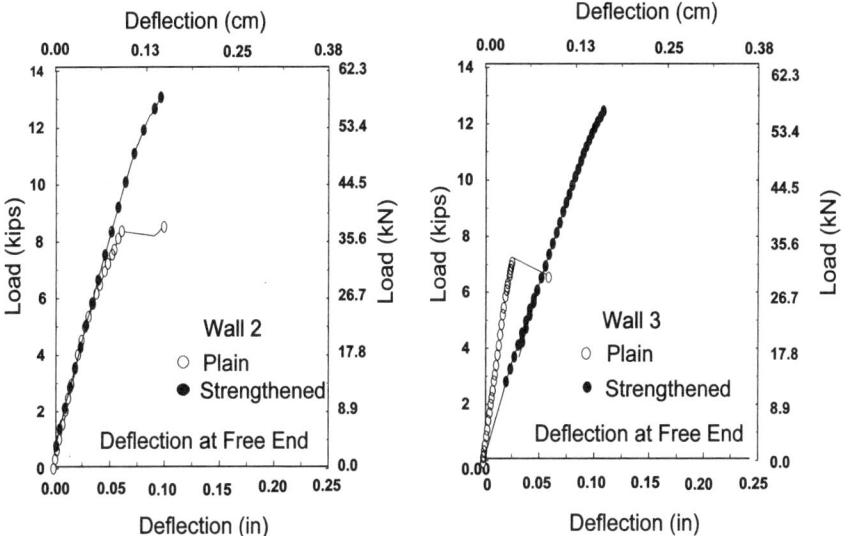

Figure 7. Load-deflection for Wall 2.   Figure 8. Load-deflection for Wall 3.

DISCUSSION

Masonry walls are subjected to in-plane bending moments and shears due to settlement loading. The bending moment is resisted by a couple formed by the tension steel in the bond beam and compression concrete in the foundation. Due to the very large lever arm, substantial, bending resistance can be realized even by lightly reinforced members. However, the absence of vertical steel, acting as shear reinforcement, causes the wall to crack. In the study, carbon fiber sheets were aligned along the length of the wall to increase the tension capacity of the bond beam. This continuous bonding also improved shear capacity of the wall since resin has a much higher shear capacity than mortar. As a result, significant strength gains were realized and indeed the second specimen, Wall 2 could not be tested to failure.

In the study, three 20 in. wide strips of carbon fiber were used and no attempt was made at material optimization. Finite element analysis carried out [8] indicated that for the walls tested, much narrower strips could be used to provide the same bending resistance. Strips as little as 8 in. or 20 in. would be adequate for most residential repairs though stiffness improvement would be poorer.

The settlement forces that develop in a wall are of course identical regardless of whether it is constructed in a seismic or non-seismic region. However, the repair scheme used in this study will need to be modified to reinforce the wall in both horizontally and vertically and provide greater ductility in seismic regions. It may be convenient to use bi-directional CFRP material for this purpose.

The repair was completed in 1995. After over four years of exposure in Florida's sub-tropical climate there was no discernible deterioration in the CFRP epoxy bond.

Figure 9. Effect of Exposure on Wall 3.

Fig. 9 compares the wall in its original to its current state. Note the corrosion in the steel that was used to anchor the ends of the CFRP sheeting, and the original sheen in the epoxy coating has lost its luster.

CONCLUSIONS

FRP material has been widely considered for the seismic retrofit of unreinforced masonry. Though devastating, earthquakes are relatively rare. In contrast, damage due to foundation settlement is much more common and affects homes nationwide. This paper presents results from an experimental study to investigate the feasibility of using uni-directional carbon fiber sheets for the repair and rehabilitation of lightly reinforced masonry walls damaged by foundation settlement.

The full-sized test walls were built by experienced masonry contractors in the same manner in which homes have been constructed over the past thirty years. Two of the walls, representative of earlier construction, were tested under simulated foundation settlement before and after strengthening. The CFRP material was applied to one surface to accommodate homeowner preference for minimizing disruption. The results of the tests are very encouraging and suggest that FRP material may be appropriate for repairing some types of foundation damage. Where damage is very severe, it may be also used in conjunction with traditional repairs for strengthening the wall after the foundation settlement has been stabilized using other means (Fig. 3). Moreover, it may be possible to optimize the repair so as to minimize costs.

The study was directed towards applications in southeastern United States where wind loading is more critical. In seismic regions, additional strengthening will be required in the horizontal and vertical directions to comply with ductility requirements for lateral-force resisting systems. In addition, the connections between the masonry walls/ roof framing /foundations will need to be appropriately strengthened. Inadequate anchorage is one of the chief causes of failure of masonry construction during earthquakes. Detailed information on the seismic retrofit of unreinforced masonry may be found elsewhere [9].

ACKNOWLEDGMENTS

The authors gratefully acknowledge the support and contribution of the Tampa Bay Chapter of the Masonry Contractors Association of Florida who donated there services and all materials needed for the fabrication of the walls. We thank Mr. Jack Harrington, P.E. for his contribution in the design of the test setup to simulate settlement. We also wish to thank Dr. Howard Kliger for technical advice and Tonen Corporation, Japan for donating materials used in the testing.

REFERENCES

1. A. B. Chance Company (1995). "Chance Helical Pier System". Bulletin 01-9501, Revised 4/95, Centralia, MO.
2. ACI 530-99 (1999). Building Code Requirements for Masonry Structures, American Concrete Institute, Farmington Hills, MI 48333.
3. ACI 318-99 (1999). Building Code Requirements for Structural Concrete, American Concrete Institute, Farmington Hills, MI 48333.
4. Hungerford, L. (1997). Brown Testing Services, Tampa, FL. Private Communication, March 27.
5. Southern Building Code Congress International (1997). Standard Building Code 1997 Edition, Birmingham, AL.
6. Southern Building Code Congress International (1993). Standard for Hurricane Resistant Residential Construction (SSTD 10-93), Birmingham, AL.
7. Hartley, A., Mullins, G. and Sen, R. (1996). "Repair of Concrete Masonry Block Walls using Carbon Fiber", *Advanced Composite Materials in Bridges and Structures* (Editor M.El-Badry), Canadian Society of Civil Engineers, Montreal, P.Q. pp. 795-802.
8. Engebretson, D., Sen, R., Mullins, G. and Hartley, A. (1996). "Strengthening Concrete Block Walls with Carbon Fiber". *Materials for the New Millennium, Proceedings of the Materials Engineering Conference Volume 2*, ASCE, New York, NY, pp. 1592-1600.
9. Federal Emergency Management Agency (1997). "NEHRP Guidelines for the Seismic Rehabilitation of Buildings", FEMA 273, and "Commentary", FEMA 274, Washington, DC, October.

Composite Retrofit of Unreinforced CMU Walls

J. GERGELY, D. T. YOUNG, J. HOOKS and N. AL-CHAAR

ABSTRACT

This study is concerned with the seismic retrofit on unreinforced hollow concrete masonry (CMU) walls using carbon fiber reinforced polymer (FRP) composites. During recent major seismic events a large number of unretrofitted masonry buildings have been damaged or collapsed. The majority of these buildings suffered complete collapse, collapse of walls, or extensive shear cracking of walls and piers. In order to investigate the possibility of using FRP composites as a retrofit material, six 4 ft by 8 ft masonry walls were tested before and after the composite retrofit. Three of the walls (shear specimens) were loaded with in-plane shear forces, and three of the walls were subjected to out-of-plane bending (bending specimens). Both of these wall types were retrofitted with three different composite laminates. A significant strength increase was observed for the repaired shear and bending specimens. Failure occurred due to excessive shear damage for the bending tests, and due to wall damage in the anchor area for the shear specimens. Since both wall faces were retrofitted with multiple composite layers, the stress level in the FRP material was well bellow its ultimate values, clearly indicating that for these tests the masonry governed the results. In addition, as a result of using an improved adhesive material, no composite delamination was observed.

INTRODUCTION

Methods for seismic rehabilitation of masonry structures include the use of shotcrete, coatings for unreinforced masonry (URM) walls, grout injection, steel bracing and stiffening elements [1]. Research on retrofit techniques for masonry elements has been conducted by Seible et al. [2]. A five-story building model was repaired successfully using carbon fiber composite materials. The dynamic characteristics of masonry bearing and shear walls have been analyzed experimentally by Al-Chaar and Hassan [3]. The glass FRP overlays enhanced the wall's shear capacity and overall strength.

University of North Carolina at Charlotte, Civil Engineering Department, 9201 University City Blvd., Charlotte, NC 28223

Composite materials have been used for the seismic rehabilitation of URM walls in the studies performed by Ehsani and Saadatmanesh [4], and by Ehsani et al. [5]. The composites were externally attached to masonry elements to increase the member's flexural and shear capacity. An experimental and analytical study was performed on masonry walls retrofitted using FRP laminates by Triantafillou [6]. Gilstrap and Dolan [7] investigated the out-of-plane bending of masonry walls reinforced with composite tapes. The performance of unreinforced masonry walls strengthened with glass and carbon composites was evaluated by Marshall et al. [8].

EXPERIMENTAL PROGRAM

The objectives of this study were to experimentally investigate precracked unreinforced hollow CMU walls retrofitted with composite laminates, subjected to in-plane shear (shear specimens) and out-of-plane bending (bending specimens) loads. As an additional test variable, the FiberBond™ laminates applied to both sides of the wall followed three different lamination schedules.

To achieve this objective, six identical 4ft X 8ft masonry walls were constructed and tested, then retrofitted and retested. 8 in. wide CMU blocks were used (the average compressive strength of the units was 1500 psi) with Type S mortar. In order to identify the walls' existing capacity, each wall was initially pretested up to the cracking point (baseline tests). Three of the walls were loaded with in-plane cyclic loads (see Figure 1a), and three of the walls were subjected to out-of-plane third point bending (see Figure 1b). The cyclic loads were applied in both directions three times, then increased to the next load step. Both of these figures also show the instrumentation used in the experiments. Displacement transducers were used to monitor the in-plane and out-of-plane deformation of the specimen at several locations. Strain gages were attached to the FRP laminates to record the stress level in the composite material throughout the specimen.

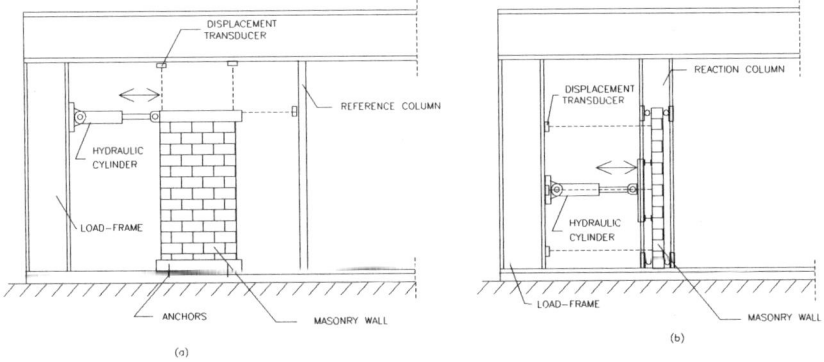

Figure 1. Test setup and instrumentation: (a) shear, and (b) bending specimens

The boundary condition for the shear test setup was a cantilever wall with fixed conditions at the bottom. The wall in the bending setup however, was pin-supported top and bottom, using a reaction column and special roller supports. In order to create a condition of low axial load in a wall system, no additional axial force was added to the loading condition.

As expected, the shear specimens developed horizontal cracks along the mortar bed joint at the walls' lower section, while the bending specimens cracked at the specimens' mid-height. The unretrofitted shear walls failed at a lateral load level of 0.8 Kips, and the bending walls failed at the average load of 1.3 Kips. Figures 2a and 2b show the load-displacement curves for the shear and bending samples. These curves indicate a linear behavior with no significant stiffness degradation. Once a crack initiated, the specimens could not sustain the applied load, and the test ended in the same load cycle.

Unreinforced masonry walls subjected to seismic forces behave in a very brittle way, and fail with little or no warning. By strengthening such a non-ductile structural element with composites, a linear material, the member strength characteristics are considered rather then its ductility properties. After each wall was carefully removed from the loading frame using a special lifting system, the surface of the specimens was prepared using a wire brush, then vacuumed. To increase the strength of the walls, FiberBond™ composite laminates were applied to both sides of the samples. Prior to this however, a thin layer of structural adhesive was applied first. This high viscosity epoxy adhesive provided an even surface for the composite application.

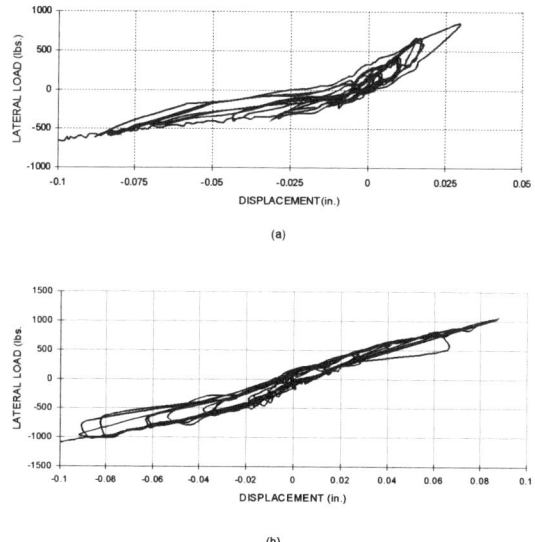

Figure 2. Baseline specimen results: (a) shear, and (b) bending tests

Factors that are considered in a masonry wall retrofit design are, among others, the height-to-thickness and height-to-width ratios, the level of axial load, and the capacity demand. Similarly to any orthotropic material, the effectiveness of the composite retrofit also depends on the orientation of the carbon/glass fibers. A [±45] layout is the most effective (although not always the most practical) to carry shear forces in a shearwall. A laminate aligned along the height of the wall, [0_2] layout, is optimal for out-of-plane bending loads. Finally, a [0/90] laminate will provide an efficient method to strengthen walls supported on all four edges and subjected to two-way bending. To identify the effectiveness of all three layouts, each of them were applied and analyzed individually. Since an exterior shear wall has to withstand seismic forces, as well as wind loads at some point in its lifetime, these walls are usually retrofitted to satisfy both loading conditions (not simultaneously). Therefore, both specimen types (shear and bending) were strengthened with a similar composite configuration.

The results of the retrofitted walls show a significant increase in both the in-plane shear and out-of-plane bending capacity. Figure 3a shows the load-deformation curve for the [±45] composite retrofit sample. Once again, these curves indicate a linear behavior, with no significant damage before failure was observed. The shear specimens reached a maximum lateral load of 9.3 Kips, a strength increase of eleven times. All these specimens lost their load carrying capacity due to extensive damage at the bottom of the walls, near the supports, well before the composite material could reach its ultimate strength. This damage in the vicinity of the anchoring system prevented the wall to fully develop its capacity. Therefore, these three shearwalls were retested in a new setup, having a height-to-weight ratio of only 0.5 (4ft X 2ft), compared to the original ratio of 2. In this new setup, the walls reached in average a surprising load level of 45 Kips. Failure occurred when the masonry crashed in compression.

The bending specimen with a [0/90] composite retrofit reached a lateral load level of 41 Kips, as shown in Figure 3b. This capacity level represents an increase of 31 times compared to the baseline specimens. The maximum lateral deflection at the wall mid-height was 0.66 in. These bending specimens behaved like a traditional composite sandwich panel, which takes advantage of the high tensile/compressive strength material at the face of the inner core (masonry, in this case). All the bending specimens developed huge shear cracks prior to failure. Therefore, once again, the (shear) capacity of the unreinforced masonry wall governed the member strength. Figure 4a shows the out-of-plane bending setup, including the loading device and reaction columns. The same specimen can be seen in Figure 4b after failure. This picture also illustrates how the composite laminates confine the crushed concrete blocks, and diminish the danger of falling debris from a damaged building.

CONCLUSIONS

The experimental results showed that the FRP laminates significantly increased the in-plane shear and out-of-plane bending capacity of precracked unreinforced hollow masonry walls. In addition, as a result of using an improved adhesive material, no composite delamination was observed. However, in all cases the shear and bearing capacity of the masonry walls governed the behavior of composite retrofitted members.

ACKNOWLEDGMENTS

The authors would like to acknowledge the financial support provided by Edge Structural Composites.

REFERENCES

1. Federal Emergency Management Agency. October 1997. *NEHRP Guidelines for the Seismic Rehabilitation of Buildings.* FEMA 273, Washington, DC.
2. Seible, F., G. Hegemier, N. Priestley, G. Kingsley, A. Igarashi, and A. Kurkchubasche. August 1990. *"Preliminary Results from the TCCMAR 5-story Full-Scale Reinforced Masonry Research Building Test,"* The Masonry Society Journal, 12(1):53-60.
3. Al-Chaar, G.K., and H.A. Hassan. May 1999. "Seismic Testing and Dynamic Analysis of Masonry Bearing and Shear Walls Retrofitted with Overlay Composite," presented at the ICE '99, Cincinnati, May 10-12, 1999.
4. Ehsani, M.R., and H. Saadatmanesh. December 1996. *"Seismic Retrofit of URM Walls with Fiber Composites,"* The Masonry Society Journal, 14(2):63-72.
5. Ehsani, M.R., H. Saadatmanesh, and A. Al-Saidy. 1997. *"Shear Behavior of URM Retrofitted with FRP Overlays,"* J. of Composites for Construction, ASCE, 1(1):17-26.
6. Triantafillou, T.C. 1998. "Strengthening of Masonry Structures Using Epoxy-Bonded FRP Laminates," J. of Composites for Construction, ASCE, 2(2):96-104.
7. Gilstrap, J.M., and C.W. Dolan. 1998. *"Out-of-plane Bending of FRP-reinforced Masonry Walls,"* Composites Science and Technology, 58:1277-1284.
8. Marshall, O.S., S.C. Sweeney, and J.C. Trovillion. 1999. "Seismic Rehabilitation of Unreinforced Masonry Walls," presented at the Fourth International Symposium on Fiber Reinforced Polymer Reinforcement for Reinforced Concrete Structures, Baltimore.

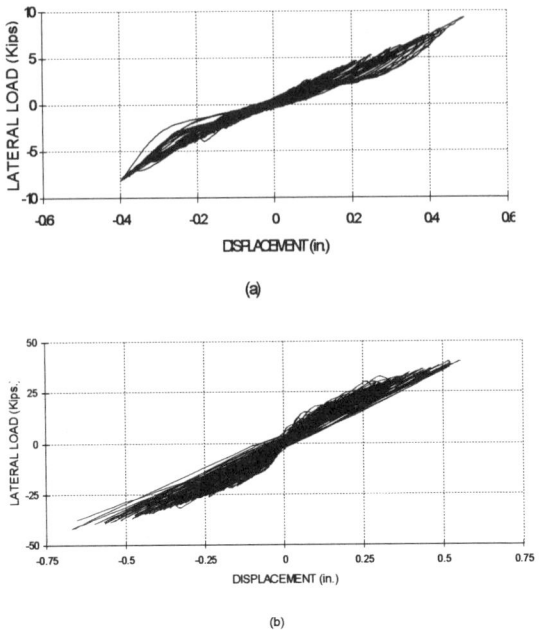

Figure 3. Repaired specimen results: (a) shear, and (b) bending tests

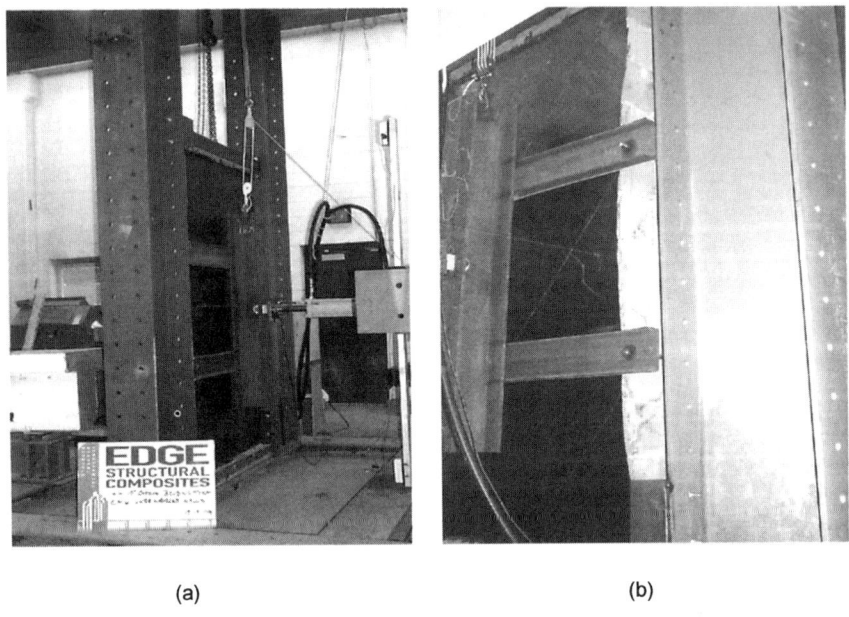

(a)   (b)

Figure 4. Out-of-plane bending specimens

A Crucial Link between Building Codes and Seismic Strengthening Technologies

B. N. HORECZKO

To understand the proper role of ICBO Evaluation Service (ICBO ES) in fostering new technologies for seismic strengthening of structures a brief introduction to the function of building codes, specifically the *Uniform Building Code* (UBC), is warranted.

The UBC, published by the International Conference of Building Officials (ICBO), is the most widely used building code in the United States. It is, in general, a performance-based document setting forth minimum fire, life-safety and structural design requirements for buildings. To permit the use of new materials and designs not covered by the UBC, the code provides a mechanism for approval of new materials and designs based on equivalency to specified code requirements. The UBC empowers the building official to analyze documentation justifying performance of new materials, products and designs. To assist the building official in this task, ICBO ES was established as a nonprofit subsidiary corporation of ICBO. Its staff of structural, civil, mechanical, fire-protection and quality engineers reviews test data, calculations and material specifications. Through such reviews, ICBO ES has published thousands of evaluation reports on building materials, products and designs. These evaluation reports are widely used by building officials, architects, engineers, designers, product specifiers and contractors. The reports form the basis for the building official's approval of new technology for building construction.

Briefly, here is how the evaluation process works using as an example reinforced plastics and composites used to strengthen structures to resist seismic forces. These materials have a good performance history in aerospace and other industries, but they have not yet attained recognition in civil and structural engineering applications as materials with established structural design values. To assist in acceptance of these products, ICBO ES helps the manufacturers provide independent information to the code official, and others, regarding the suitability of composites in building-code

---

Senior Staff Engineer, ICBO Evaluation Services, Inc, 5360 Workman Hill Road, Whittier, CA 90601, USA

applications; this is accomplished by issuing technical evaluation reports on these materials.

Before a technical evaluation report can be written, standards must be developed for their evaluation. These standards or instruments for product evaluation are known as ICBO ES acceptance criteria. For reinforced plastics and composites, they would be developed by ICBO ES staff working in conjunction with manufacturers, industry groups and experts in the composites field. The proposed criteria would be discussed in public hearings presided over by the ICBO ES Evaluation Committee. All interested parties known by ICBO ES to have expertise in composite technology would be invited to attend the hearings and comment on the proposed criteria. Their comments would be given consideration in the development of the final draft of the criteria.

One or more of these public hearings may be necessary to refine the acceptance criteria. Upon approval by the Evaluation Committee, the criteria serves as the basis for testing to establish the integrity and suitability of composite materials for their intended use. Upon successful compliance with the testing requirements of the criteria, and review and approval of ICBO ES staff, an evaluation report is issued. The evaluation report describes the product, its installation requirements, and its design and conditions of use under provisions of the UBC. The evaluation reports are used by code officials to help them approve products for code compliance. An example of an ICBO ES acceptance criteria (AC125) is included as part of this paper (Appendix A).

A significant development germane to this process is the publication of the *International Building Code®* (IBC). Scheduled for publication in February 2000, the IBC is considered by ICBO to be the successor to the UBC, and is intended for national and international applications. ICBO ES is already using this document as source material when issuing evaluation reports on reinforced plastics and composites, if specifically requested by report applicants.

The above-noted process is a brief description of how ICBO ES is involved in the use of new products and designs as alternates to conventional products and designs described in the UBC. In this way, ICBO ES functions as a technology transfer organization, facilitating the transition of new materials and products from research and development to production and commercial end use.

APPENDIX A: ICBO ES acceptance criteria (AC125) Document (see next page).

ICBO Evaluation Service, Inc.

A subsidiary corporation of the International Conference of Building Officials

5360 WORKMAN MILL ROAD • WHITTIER, CALIFORNIA 90601-2299 • (562) 699-0541
FAX (562) 695-4694

ACCEPTANCE CRITERIA FOR CONCRETE AND REINFORCED AND UNREINFORCED MASONRY STRENGTHENING USING FIBER-REINFORCED, COMPOSITE SYSTEMS

AC125

April 1997

PREFACE

Evaluation reports issued by the ICBO Evaluation Service, Inc. (ICBO ES), are based upon performance features of the *Uniform Building Code*™, ICBO *Uniform Mechanical Code*™ and related codes. Section 104.2.8 of the Uniform Building Code is the primary charging section upon which evaluation reports are issued. Section 104.2.8 reads as follows:

The provisions of this code are not intended to prevent the use of any material, alternate design or method of construction not specifically prescribed by this code, provided any alternate has been approved and its use authorized by the building official.

The building official may approve any such alternate, provided the building official finds that the proposed design is satisfactory and complies with the provisions of this code and that the material, method or work offered is, for the purpose intended, at least the equivalent of that prescribed in this code in suitability, strength, effectiveness, fire resistance, durability, safety and sanitation.

The building official shall require that sufficient evidence or proof be submitted to substantiate any claims that may be made regarding its use. The details of any action granting approval of an alternate shall be recorded and entered in the files of the code enforcement agency.

The attached acceptance criteria for the general code sections noted have been issued to provide all interested parties with guidelines on implementing performance features of the codes. The attached acceptance criteria were developed and adopted following public hearings conducted by the Evaluation Committee. These criteria may be revised from time to time as the need dictates.

ICBO ES may consider alternate criteria, provided the proponent submits valid data demonstrating that the alternate criteria are at least equivalent to the attached criteria and otherwise meet the applicable performance requirements of the codes. Notwithstanding that a material, type or method of construction, or equipment, meets the attached acceptance criteria, or it can be demonstrated that valid alternate criteria are equivalent and otherwise meet the applicable performance requirements of the codes, if the material, product, system or equipment is such that either unusual care with its installation or use must be exercised for satisfactory performance, or malfunctioning is apt to cause unreasonable property damage or personal injury or sickness relative to the benefits to be achieved by the use thereof, ICBO ES retains the right to refuse to issue or renew an evaluation report.

Published by the

International Conference of Building Officials
5360 WORKMAN MILL ROAD • WHITTIER, CALIFORNIA 90601-2298

Copyright © 1997 ICBO Evaluation Service, Inc.

PRINTED IN THE U.S.A.

ACCEPTANCE CRITERIA FOR CONCRETE AND REINFORCED AND UNREINFORCED MASONRY STRENGTHENING USING FIBER-REINFORCED, COMPOSITE SYSTEMS

1. INTRODUCTION
2. DEFINITIONS 3
   - 2.1 Design Values 3
   - 2.2 Composite Material 3
   - 2.3 Cracking Load and Displacement 3
   - 2.4 Yielding Load and Displacement 3
   - 2.5 Passive and Active Composite Systems 3
3. REQUIRED INFORMATION 3
   - 3.1 Description 3
   - 3.2 Installation Instructions 3
   - 3.3 Structural Design 3
4. TESTING LABORATORIES AND REPORTS OF TESTS 4
5. QUALIFICATION TESTS 4
   - 5.1 Qualification Test Plan 4
   - 5.2 Columns 4
     - 5.2.1 Flexural Tests 4
       - 5.2.1.1 Configuration 4
       - 5.2.1.2 Procedure 4
     - 5.2.2 Shear Tests 4
       - 5.2.2.1 Configuration 4
       - 5.2.2.2 Procedure 4
   - 5.3 Beam-to-Column Joints 4
     - 5.3.1 Configuration 4
     - 5.3.2 Procedure 4
   - 5.4 Beams 4
     - 5.4.1 Flexural Tests ....
       - 5.4.1.1 Configuration ....
       - 5.4.1.2 Procedure 4
     - 5.4.2 Shear Tests 4
       - 5.4.2.1 Configuration 4
       - 5.4.2.2 Procedure 4
   - 5.5 Walls 4
     - 5.5.1 Wall Flexural Tests (Out-of-Plane Load) 4
       - 5.5.1.1 Configuration 4
       - 5.5.1.2 Procedure 4
     - 5.5.2 Wall Shear Tests (In-Plane Shear) 4
       - 5.5.2.1 Configuration 4
       - 5.5.2.2 Procedure 4
   - 5.6 Wall-to-Floor Joints 4
     - 5.6.1 Configurations 4
     - 5.6.2 Procedure 4
   - 5.7 Slabs (Flexural Tests) 5
     - 5.7.1 Configuration 5
     - 5.7.2 Procedure 5
   - 5.8 Physical and Mechanical Properties of Composite Materials 5
     - 5.9.1 Procedure 5
     - 5.9.2 Conditions of Acceptance 5
   - 5.10 Freezing and Thawing 5
     - 5.10.1 Procedure 5
     - 5.10.2 Conditions of Acceptance 5
   - 5.11 Aging 5
     - 5.11.1 Procedure 5
     - 5.11.2 Conditions of Acceptance 5
   - 5.12 Alkali Soil Resistance 5
     - 5.12.1 Procedure 5

		5.12.2 Conditions of Acceptance	5
	5.13	Fire-resistant Construction	5
	5.14	Interior Finish	5
	5.15	Fuel Resistance	5
	5.16	Adhesive Lap Strength	5
	5.17	Bond Strength	5
6.	QUALITY CONTROL		5
	6.1	Manufacturing	5
	6.2	Installation	5
7.	FINAL SUBMITTAL		5
	7.2	Test Report	5
	7.3	Design Criteria	5
		7.3.1 Design Criteria Report	5
		7.3.2 Minimum Acceptable Design Criteria	6
		7.3.2.1 Flexural Strength Enhancement	6
		7.3.2.2 Bond Strength of Fiber to Concrete or Masonry	6
		7.3.2.3 Axial Load Capacity Enhancement	6
		7.3.2.3.1 Longitudinal Fiber	6
		7.3.2.3.2 Transverse Fiber	6
		7.3.2.3.2.1 Circular Sections	6
		7.3.2.3.2.2 Rectangular Sections	6
		7.3.2.4 Ductility Enhancement	6
		7.3.2.4.1 Circular Sections	6
		7.3.2.4.2 Rectangular Sections	6
		7.3.2.5 Lap-Splice Confinement	6
		7.3.2.6 Shear Strength Enhancement	7
		7.3.2.6.1 Circular Sections	7
		7.3.2.6.2 Rectangular Beam or Column Sections	7
		7.3.2.6.3 Rectangular Wall Sections	7
		7.3.2.6.4 Shear Strength Reduction Factor	7
		7.3.2.7 Enhancement Using Active Composite Systems	7
	7.4	Quality Control	7
	7.5	Nomenclature	7

1. INTRODUCTION

These criteria establish minimum requirements for the issuance of ICBO Evaluation Service, Inc. (ICBO ES), evaluation reports on fiber-reinforced composite systems used to strengthen concrete and masonry structural elements. These reports consider these systems as alternates to those covered in the Uniform Building Code (hereafter cited as "the code").

2. DEFINITIONS

2.1 **Design Values:** The composite material's load and deformation design capacities, that are based on either working stress or ultimate strength methods.

2.2 **Composite Material:** A combination of high-strength fibers and polymer matrix material. This composite may be applied either during manufacture of the structural element or at the project location.

2.3 **Cracking Load and Displacement:** Load and displacement at which the moment-curvature relationship of the concrete or masonry section first changes slope or at which the cracking moment is reached.

2.4 **Yielding Load and Displacement:** Load and displacement at which longitudinal steel reinforcement of the concrete or masonry section first yields.

2.5 **Passive and Active Composite Systems:** Active systems are those composite systems where the composite materials are post-tensioned after installation by means such as pressure injection be-tween the composite material and the concrete or masonry section. Passive systems are not post-tensioned after installation.

3. REQUIRED INFORMATION

3.1 **Description:** A detailed description of the strengthening system is needed, including the following items:

1. Description and identification of the product or system.
2. Restrictions or limitations on use.

3.2 **Installation Instructions:** Instructions shall include the following items:

1. Description of how the product or system will be used or installed in the field.
2. Procedures establishing quality control in field installation.
3. Requirements for product handling and storage.
4. Fastener installation into structural elements.
5. For systems that depend on bond between the system and the substrate, on-site testing of bond to the substrate is required.

3.3 **Structural Design:** The structural applications of the system shall address the following items:

1. Clarification of recognition under either Chapter 19 or Chapter 21 of the code.
2. Complete description of details.
3. Details on how the product or system does or does not comply with Chapter 19 of the code, including conformities and deviations. Details should include positive statements that the product or system does comply with Chapter 19 or 21 in the following areas . . . ; and negative statements that it does not comply in the following areas. . . .

4. Details and examples of how the product or system is designed and analyzed, including formulas, with procedures and properties needed for design and analysis. The engineering analysis should define failure modes or force and deflection limit states.

5. Use of anchors shall be considered where composite material bond to substrate is critical.

4. TESTING LABORATORIES AND REPORTS OF TESTS

4.1 Testing laboratories shall comply with the ICBO ES Acceptance Criteria for Laboratory Accreditation (AC89).

4.2 Test reports shall comply with the ICBO ES Acceptance Criteria for Test Reports (AC85).

5. QUALIFICATION TESTS

5.1 Qualification Test Plan: The intent of testing is to verify the design equations and assumptions used in the engineering analysis. All or part of the tests described in this section, and any additional tests identified for special features of the product or system, shall be specified. The test plan shall be a complete document.

Overall, qualification testing must provide data on material properties, force and deformation limit states, and failure modes, to support a rational analysis procedure. The specimens shall be constructed under conditions specified by the manufacturer, including curing. Tests must simulate the anticipated loading conditions, load levels, deflections, and ductilities.

5.2 Columns:

5.2.1 Flexural Tests:

5.2.1.1 Configuration: Column specimens shall be configured to induce flexural limit states or failure modes. Either cantilever or double fixity (reverse curvature) is permitted in specimens. Extremes of dimensional, reinforcing, and strength parameters shall be considered.

5.2.1.2 Procedure: For seismic or wind-load applications, the lateral load procedure shall conform to Figure 1. For gravity (nondynamic) loading applications, the load may be monotonically applied. Axial loads within a specific range shall be applied. The limit states shall be determined based on material properties and an extreme concrete or masonry fiber compression strain of 0.003.

5.2.2 Shear Tests:

5.2.2.1 Configuration: Column specimen spans shall be configured to induce shear limit states or failure modes. Double fixity (reverse curvature) is required. Extremes of dimensional, reinforcing, and compressive strength parameters shall be considered.

5.2.2.2 Procedure: For seismic or wind-load application, the lateral load procedure shall conform to Figure 1. For gravity (nondynamic) loading application, the load may be monotonically applied. Axial loads within a specific range shall be applied. The limit states shall be determined based on material properties.

5.3 Beam-to-Column Joints:

5.3.1 Configuration: The beam-to-column joint shall be configured to induce joint-related limit states or failure modes. The column portion may be constructed to represent a section between inflection points. Extremes of dimensional, reinforcing and compressive strength parameters shall be considered.

5.3.2 Procedure: The lateral load procedure shall conform to Figure 1. A vertical load shall be continuously applied and varied within a specified range. The limit states shall be determined based on material properties.

5.4 Beams:

5.4.1 Flexural Tests:

5.4.1.1 Configuration: Beam spans shall be configured to induce flexural limit states or failure mode. Either simple or rigid supports are permitted. Extremes of dimensional, reinforcing, and compressive strength parameters shall be covered.

5.4.1.2 Procedure: For seismic or wind-load applications the lateral load procedure shall conform to Figure 1. For gravity (nondynamic) loading application, the load may be monotonically applied. The limit states shall be determined based on material properties and an extreme concrete or masonry fiber compression strain of 0.003.

5.4.2 Shear Tests:

5.4.2.1 Configuration: Beam spans shall be configured to induce shear limit states or failure modes. Either simple or rigid supports are permitted. Extremes of dimensional, reinforcing, and compressive strength parameters shall be considered.

5.4.2.2 Procedure: For seismic or wind-load application, the lateral load procedure shall conform to Figure 1. For gravity loading, the load may be monotonically applied. The limit states shall be determined based on material properties.

5.5 Walls:

5.5.1 Wall Flexural Tests (Out-of-Plane Load):

5.5.1.1 Configuration: Wall flexural specimens shall be configured to induce out-of-plane flexural limit states and failure modes. Extremes of dimensional, reinforcing, and compressive strength parameters shall be considered.

5.5.1.2 Procedure: Specimens may be axially loaded to consider effects of axial loads. The loading in the out-of-plane direction may be applied at third-points, by air-bags or by other means representing actual conditions. The lateral load procedure consists of:

5.5.1.2.1 Load specimens in both directions, to find cracking and yielding load and deformation at first cracking. For unreinforced masonry, only cracking load and deformation are required.

5.5.1.2.2 At least two cycles of loading in both directions, under displacement control at each deformation level. The deformation levels shall consist of multiples of the deformation at yielding for reinforced concrete or masonry sections or cracking for unreinforced masonry sections.

5.5.1.2.3 The specimens are loaded in both directions until the strengthening system is damaged, its capacity is reached, or desired limit states are achieved.

5.5.2 Wall Shear Tests (In-Plane Shear):

5.5.2.1 Configuration: Wall specimens shall be configured to induce in-plane shear limit states or failure modes. Extremes of dimensional, reinforcing and compressive strength parameters shall be considered.

5.5.2.2 Procedure: Specimens may by axially loaded to consider effects of axial loads. The lateral load procedure consists of:

5.5.2.2.1 Load specimens in both directions to find cracking and yielding load and deformation. For unreinforced masonry, only cracking load and deformation are required.

5.5.2.2.2 The specimens are loaded in both directions until the strengthening system is damaged, its capacity is reached, or desired limit states are achieved.

5.6 Wall-to-Floor Joints:

5.6.1 Configurations: The specimens shall be configured to induce joint-related limit states or failure modes. Extremes of dimensional, reinforcing and compressive strength parameters shall be considered.

5.6.2 Procedure: For seismic or wind-loading applications the lateral load procedure shall conform to Figure 1. For gravity load applications, the load may be monotonically applied. The vertical load shall be applied to floors. The limit states shall be determined based on material properties.

5.7 Slabs (Flexural Tests):

5.7.1 Configuration: Slab spans shall be configured to include flexural limit states or failure modes. Either simple or rigid supports are permitted. Extremes of dimensional, reinforcing and compressive strength shall be considered.

5.7.2 Procedure: For seismic or wind-load applications, the lateral load procedure shall conform to Figure 1. For gravity (nondynamic) loading application, the load may be monotonically applied. The limit states shall be determined based on material properties and an extreme concrete fiber compression strain of 0.003.

5.8 Physical and Mechanical Properties of Composite Materials: Required physical and mechanical properties are shown in Table 1. These properties, including creep, CTE and impact, shall be considered in the design criteria and limitations.

5.9 Exterior Exposure:

5.9.1 Procedure: Structural composite materials are tested according to ASTM G 23. Six specimens, measuring $^3/_4$ inch by 10 inches (19.1 by 254 mm), are required. These specimens also may be cut from a panel that has been coated and painted to represent end-use conditions. Five specimens are exposed to cycles consisting of 102 minutes light and 18 minutes light and water spray in the weatherometer chamber. Minimum duration is 2,000 hours. The black-body temperature is 145°F. Both exposed and control specimens are then tested to ASTM D 3039, for tensile strength, tensile modulus and elongation. Five other specimens are controlled samples.

5.9.2 Conditions of Acceptance: Control and exposed specimens are visually examined using 5x magnification. Surface changes affecting performance, such as erosion, cracking, crazing, checking, and chalks, are subject to further investigation. The specimens shall retain at least 90 percent of tensile properties generated on control specimens.

5.10 Freezing and Thawing:

5.10.1 Procedure: Fifteen samples are conditioned in a 100 percent relative humidity chamber at 100°F for three weeks. Each cycle is 4 hours, minimum, in a 0°F freezer followed by 12 hours, minimum, in the humidity chamber. At least twenty cycles are required.

Control specimens and cycled specimens are then tested according to Table 1 for tensile strength, tensile modulus, elongation, glass transition temperature, and interlaminar shear strength. Specimens are tested in the primary direction.

5.10.2 Conditions of Acceptance: Control specimens and cycled specimens are visually examined using 5x magnification. Surface changes affecting performance, such as erosion, cracking, crazing, checking and chalking, are unacceptable. The cycled specimens shall retain at least 90 percent of the tensile properties determined for conditioned specimens.

5.11 Aging: These tests shall be considered in design criteria and limitations.

5.11.1 Procedure: Both wet and dry specimens are aged according to Table 2. Both exposed and control specimens are then tested to Table 1 for tensile strength, tensile modulus, elongation, glass transition temperature, and interlaminar shear strength. Specimens are tested in the primary direction. Five specimens per condition are required.

5.11.2 Conditions of Acceptance: Control and exposed specimens are visually examined using 5x magnification. Surface changes affecting performance, such as erosion, cracking, crazing, checking, and chalking, are unacceptable. The exposed specimens shall retain the percentage of tensile properties generated on conditioned specimens noted in Table 2.

5.12 Alkali Soil Resistance:

5.12.1 Procedure: Tests are done on five specimens according to ASTM D 3083, Section 9.5, for 1,000 hours. Both conditioned and exposed specimens are then tested for tensile strength, tensile modulus, and elongation according to ASTM D 3039.

5.12.2 Conditions of Acceptance: Conditioned and exposed specimens are visually examined using 5x magnification. Surface changes, such as erosion, cracking, crazing, checking, and chalking, are unacceptable. The exposed specimens shall retain at least 90 percent of tensile properties generated on conditioned specimens.

5.13 Fire-resistant Construction: The effect of the fiber-reinforced, composite system on fire-resistance construction shall be evaluated according to Section 703 of the code.

5.14 Interior Finish: The classification of the fiber-reinforced, composite system as an interior finish shall be determined according to Section 802 of the code.

5.15 Fuel Resistance: Tested specimens are tested according to ASTM C 581. The specimens are exposed to diesel fuel reagent for 4 hours, minimum. Specimens are tested according to Table 1 for tensile strength, tensile modulus, elongation, glass transition temperature, and interlaminar shear strength.

5.16 Adhesive Lap Strength: This test applies to prefabricated systems. Specimens of the adhesive are tested according to ASTM D 3165 for exposures in Table 2, and Sections 5.10 and 5.15.

5.17 Bond Strength: The test applies to systems that bond to the substrate. Tests are conducted for tension according to ASTM C 297 where the composite material bonds two substrate elements together, and for shear using a method acceptable to ICBO ES staff. Specimens are exposed according to Table 2 and Section 5.10.

6. QUALITY CONTROL

6.1 Manufacturing: Quality assurance procedures during manufacture of the system components shall be described in a quality control manual complying with the ICBO ES Acceptance Criteria for Quality Control Manuals (AC10).

6.2 Installation: All installations shall be done by applicators approved by the proponent of the system. The quality assurance program shall be documented. Special inspection is required and shall comply with Section 1701 and other applicable sections of the code. Duties of the special inspector shall be described and included in the evaluation report. The maximum debonded area permitted after installation of bonded systems is 2 square inches (1290.32 mm^2).

7. FINAL SUBMITTAL

7.1 The final submittal will consist of a test report or test reports, and a design criteria report, as described in this section. The final submittal shall include the data described in Section 3 of this criteria. Contents of the final submittal are described in the following subsections:

7.2 Test Report: The independent laboratory shall report on the qualification testing performed according to the approved test plan. Besides the information requested in Section 4, the test report must include the following:

1. Information noted in the reference standard.
2. Description of test setup.
3. Rate and method of loading.
4. Deformation and strain measurements.
5. Modes of failure.
6. Strain measurements.

7.3 Design Criteria

7.3.1 Design Criteria Report: The report shall include a complete analysis and interpretation of the qualification test results. Design stress and strain criteria for concrete and reinforced and unreinforced masonry systems shall be specified based on the analyses, but shall not be higher than specified in Section 7.3.2.

Design stresses and strains shall be based on a characteristic value approach verified by test data. The CTE, creep and impact values determined in Table 1, Section 5, shall be considered in the design procedure. The design shall consider secondary stresses resulting when dead loads are relieved during application and subsequently reapplied. Adoption of the minimum acceptable standards for design outlined in Section 7.3.2 does not eliminate the need for structural testing.

Situations not covered in Section 7.3.2 shall be subject to special considerations and testing, and design values should be compatible with the conservative approach adopted in Section 7.3.2, and discussed in reference [1].

7.3.2 Minimum Acceptable Design Criteria:

7.3.2.1 Flexural Strength Enhancement: Fiber-reinforced composite material bonded to surfaces of concrete or masonry may be used to enhance the design flexural strength of sections by acting as additional tension or compression reinforcement. In such cases, section analysis shall be based on normal assumptions of strain compatibility between concrete, reinforcement and composite material. The enhancement of axial force provided by a fiber element of effective thickness t_{jf} oriented at angle θ to the direction of member axis, shall be

$$\Delta F = \frac{t_f \cos^2 \theta f_{jf}}{\text{unit width}} \quad (1)$$

where $f_{jf} = E_f \epsilon_f \cos^2 \theta \le 0.75 f_{uj}$ and ϵ_f is the strain in the concrete or masonry to which the fiber is bonded at the section strength in the direction of the member axis. Unless the compression zone is confined by transversely-oriented fiber outside the flexural fiber, an extreme compression strain of $\epsilon_c = 0.003$ shall be assumed in determining flexural strength. If $\theta > 45°$, the fiber contribution to flexural strength shall be ignored unless equal fiber quantities are provided with an orientation of θ to the member axis.

Dependable flexural strengths shall be determined by multiplying the nominal flexural strength, including the effects of fiber according to Equation (1), by the appropriate flexural strength reduction factor according to the UBC.

7.3.2.2 Bond Strength of Fiber to Concrete or Masonry: Where the performance of the composite material depends on bond, the bond strength of fiber-reinforced composite material to concrete or masonry shall not be less than the characteristic flexural tension capacity f_t of the concrete or masonry. Under ultimate flexural strength conditions, bond stress between fiber-reinforced composite material and concrete or masonry shall not exceed

$$u_a = \frac{d(t_f f_j)}{dx} \le 0.75 f_t \quad (2)$$

where x is the direction parallel to the fiber. Equation (2) should be evaluated at sections where rate of change in fiber net force $t_f f_j$ is a maximum. This will normally correspond to locations of maximum shear force.

7.3.2.3 Axial Load Capacity Enhancement: Fiber-reinforced composite material may be bonded to external surfaces of concrete or masonry members to enhance axial load capacity. Depending on the section shape, axial load capacity enhancement may be provided by longitudinal and/or transverse orientation of the fiber.

7.3.2.3.1 Longitudinal Fiber: All sections may have axial load capacity enhanced by fibers with a significant component of the angle parallel to member axes. In such cases, the principles stated in Section 7.3.2.1 shall apply with strength enhancement following Equation (1). Unless the section is effectively confined by transverse fiber with angle $\theta > 75°$ to the member axis, outside the longitudinal fiber, the enhancement of axial strength, given by Equation (1), shall apply at a fiber strain $\epsilon_f = 0.002$. Where the section is effectively confined, a higher compression strain $\epsilon_f = \epsilon_{cu}$ may be used, where ϵ_{cu} is given by Equation (8) or (9).

7.3.2.3.2 Transverse Fiber: Circular sections, and rectangular sections where the ratio of longer to shorter section side dimension is not greater than 1.5, may have axial compression capacity enhanced by the confining effect of fiber-reinforced composite material placed with fibers running essentially perpendicular to the members' axis $\theta \ge 75°$.

7.3.2.3.2.1 Circular Sections: Compression strength f'_{cc} of concrete of circular columns, diameter D, with fiber of effective thickness t_f at angle $\theta \ge 75°$ to the longitudinal axis of the member, shall be given by

$$f'_{cc} = f_c \left[2.25 \sqrt{1 + 7.9 \frac{f'_l}{f'_c}} - 2 \frac{f'_l}{f'_c} - 1.25 \right] \quad (3)$$

where

$$f_l = 0.26 \rho_{sj} f_{uj} \sin^2 \theta \quad (4)$$

and

$$\rho_{sj} = \frac{4 t_f}{D} \quad (5)$$

7.3.2.3.2.2 Rectangular Sections: Compression strength f'_{cc} of concrete in rectangular columns of side lengths B and H where $B \le H \le 1.5B$, and with fiber of effective thickness t_f at angle $\theta \le 45°$ to the longitudinal axis of the member, shall be given by

$$f'_{cc} = f'_c (1 + 1.5 \rho_s \cos^2 \theta) \quad (6)$$

where

$$\rho_{sj} = 2 t_f \frac{(B + H)}{BH} \quad (7)$$

For rectangular sections confined with transverse fiber-reinforced composite material, section corners must be rounded to a radius not less than $^3/_4$ inch (20 mm) before placing composite material. Axial compression capacity enhancement by fiber-reinforced composite material to rectangular sections within aspect ratio $H/_B \ge 1.5$ shall be subject to special analysis confirmed by test results.

7.3.2.4 Ductility Enhancement: Fiber-reinforced composite material oriented essentially transversely to the members axis may be used to enhance flexural ductility capacity of circular and rectangular sections where the ratio of longer to shorter section dimension does not exceed 1.5. The enhancement is provided by increasing the effective ultimate compression strain of the section.

7.3.2.4.1 Circular Sections: Ultimate compression strain of circular sections of diameter D, confined with fiber of effective thickness t_f at angle $\theta = 90°$ to the longitudinal axis of the member, shall be given by

$$\epsilon_{cu} = 0.004 + \frac{2.5 \rho_{sj} f_{uj} \epsilon_{uj}}{f'_{cc}} \quad (8)$$

where f'_{cc} is given by Equation (3), and ρ_{sj} by Equation (5).

7.3.2.4.2 Rectangular Sections: Ultimate compression strains of rectangular sections of side lengths B and H where $H \le 1.5B$, and with fiber of effective thickness t_f at an angle θ to the longitudinal axis of the member, shall be given by

$$\epsilon_{cu} = 0.004 + \frac{12.5 \rho_{sj} f_{uj} \epsilon_{uj}}{f'_{cc}} \quad (9)$$

where f'_{cc} is given by Equation (6) and ρ_{sj} by Equation (7).

For rectangular sections confined with transverse fiber-reinforced composite material, section corners must be rounded to a radius of not less than $^3/_4$ inch (20 mm) before placing composite material. Ductility enhancement according to Equation (9) should not be relied on for slender members where the aspect ratio $M/VB \ge 3$.

7.3.2.5 Lap-Splice Confinement: Lap-splices in circular columns can be confined by jackets to prevent bond failure. The required volumetric ratio of fiber-reinforced composite material, at an angle θ to the longitudinal axis of the member, given by Equation (5), shall be not less than

$$\rho_{\eta} = \frac{1.4 A_b f_s}{p \ell_s f_j} \quad (10)$$

where p is the perimeter of the crack surface forming before splice failure given by the lesser of Equations (11) and (12):

$$p = \frac{\pi D'}{2n} + 2(d_b + c) \quad (11)$$

$$p = 2\sqrt{2} \ (c + d_b) \quad (12)$$

In Equation (10), the circular section is reinforced with n bars each of diameter d_b, area A_b, uniformly distributed around the section on core diameter D'. Required stress to be transferred is f_s, and the splice length ℓ_s must not be less than

$$\ell_s = \frac{0.025 d_b f_y}{\sqrt{f'_c}} \quad (13)$$

The jacket stress f_j in Equation (10) shall not be taken larger than $f_j = 0.0015 E_j \le 0.75 f_{uj}$.

Note: Rectangular sections cannot generally be effectively confined by rectangular jackets against splice failure, and so no provisions are included here.

7.3.2.6 Shear Strength Enhancement: Shear strength of circular and rectangular sections can be enhanced by fiber-reinforced composite materials with fiber oriented essentially perpendicular to the members' axis.

7.3.2.6.1 Circular Sections: Nominal shear strength enhancement for circular sections of diameter D, with fiber thickness t_f at an angle θ to the members' axis, shall be given by

$$V_{sj} = 2.25 t_f f_j D \sin^2 \theta \quad (14)$$

where

$$f_j = 0.004 E_j \le 0.75 f_{uj} \quad (15)$$

7.3.2.6.2 Rectangular Beam or Column Sections: Nominal shear strength enhancement for rectangular sections or depth H parallel to the direction of applied shear force, with fiber thickness t_f at an angle $\theta \ge 75°$ to the members' axis, shall be given by

$$V_{sj} = 2.86 t_f f_j H \sin^2 \theta \quad (16)$$

where f_j is given by Equation (15).

For rectangular sections with shear enhancement provided by transverse fiber-reinforced composite material, section corners must be rounded to a radius not less than $3/4$ inch (20 mm) before placement of the composite material.

7.3.2.6.3 Rectangular Wall Sections: Nominal shear strength enhancement for rectangular wall sections of depth H parallel to the direction of applied shear force, with fiber thickness t_f on both sides of the wall at an angle θ to the members' axis, shall be given by

$$V_{sj} = 2 t_f f_j H \sin^2 \theta \quad (17)$$

where f_j is given by Equation (15).

Where wall sections have fiber bonded to one side only at an angle $\ge 75°$ to the member axis and with anchorage provided by bonding to the wall ends, nominal shear strength enhancement shall be taken as

$$V_{sj} = 0.75 t_f f_j H \sin^2 \theta \quad (18)$$

where f_j is given by Equation (15).

7.3.2.6.4 Shear Strength Reduction Factor: Dependable shear strength enhancement shall be found by multiplying the nominal-shear strength given by Equations (14), (16), (17), or (18), as appropriate, by a shear strength reduction factor.

Note: These provisions do not apply to shear strength enhancement provided by fiber that does not extend the full section width bonding to perpendicular faces (section ends). These provisions do not apply to shear strength enhancement for flanged sections requiring placement of fiber around re-entrant corners. These cases must be subject to special study. The use of special anchors attaching the fiber-reinforced composite material at the wall edges may be effective in transferring the design fiber stress between wall or beam and fiber.

7.3.2.7 Enhancement Using Active Composite Systems: Active composite systems can be used to reduce the thickness of the composite materials required and provide active confinement for the structural member. In such cases, analysis shall be based on assumptions verified by test results, including confining pressure, confining strain, creep and other durability considerations.

7.4 Quality Control: The quality control documents described in Sections 6.1 and 6.2 shall be submitted.

7.5 Nomenclature:

d_b = reinforcement bar diameter, inches.
B = width of compressive face of a rectangular column, inches.
D = diameter of circular columns, inches.
E_j = modulus of elasticity of composite material, psi.
ΔF = increase in axial force, lb.
f'_t = tensile strength of concrete or masonry, psi.
f_ℓ = lateral confining stress, psi.
f'_{cc} = compressive strength of columns, psi.
f_{lf} = confining strength of composite material, psi.
f_{uj} = ultimate tensile strength of composite material, psi.
f_j = hoop stress developed in jacket material, psi.
H = side length of a rectangular column, inches.
ℓ_s = reinforcement bar splice length, inches.
P = perimeter of cracked surface, inches.
P_{sj} = volumetric ratio of retrofit jacket.
t_f = effective composite material thickness.
μ = displacement ductility level, defined relative to yield or cracking displacement.
M_u = bond strength between composite material and concrete or masonry, psi.
V_{sj} = shear strength enhancement provided by composite material, lb.
ε_c = concrete compression strain.
ε_{cu} = ultimate compression strain.
ε_f = strain composite material at designated strength.
ε_{cc} = strain at peak stress for confined concrete.
θ = angle of fiber direction to member axis.

Reference

Priestley, M.J. Nigel, Frieder Seible and Michele Calvi, *Seismic Design and Retrofit of Bridges* (Chapters 1 through 8). John Wiley and Sons, Inc., New York, September 1995, 672 pp.

TABLE 1—PHYSICAL PROPERTIES

PROPERTIES	TEST METHOD	NO. OF SPECIMENS[1]
Tensile strength	ASTM D 3039	20[2]
Elongation	ASTM D 3039	
Tensile modulus	ASTM D 3039	
Coefficient of thermal expansion (CTE)	ASTM D 696 or E 1142	5[2]
Creep	ASTM D 2990[3]	5[2]
Void content	ASTM D 2584[5] or D 3171[5]	5
Glass transition (Tg) temperature	ASTM D 4065	20[6]
Impact	ASTM D 3029, Method I[4]	5
Composite interlaminar shear strength	ASTM D 2344	20

[1]Specimen sets shall exhibit a coefficient of variation (COV) of 6 percent or less. Outliers are subject to further investigation according to ASTM E 178. If the COV exceeds 6 percent, the numbered specimens shall be doubled.
[2]Values shall be determined in the primary and cross (90°) directions.
[3]Test duration is 3,000 hours, minimum.
[4]Impact head is 0.625 inch. Specimens may be rectangular, measuring 4 inches by 6 inches, and are placed on 3-inch-by-5-inch supports. 250 lb-in at 0.10 inch thick is the minimum requirement.
[5]Maximum void content by volume is 6 percent.
[6]Minimum 140°F Tg is required for control and exposed specimens.

TABLE 2—ENVIRONMENTAL DURABILITY TEST MATRIX

ENVIRONMENTAL DURABILITY TEST	RELEVANT SPECIFICATIONS	TEST CONDITIONS	TEST DURATION	PERCENT RETENTION Hours	
				1,000	3,000
Water resistance	ASTM D 2247 ASTM E 104	100 percent, 100 ± 2° F	1,000, 3,000 and 10,000 hours		
Saltwater resistance	ASTM D 1141 ASTM C 581	Immersion at 73 ± 2° F	1,000, 3,000 and 10,000 hours	90	85
Alkali resistance	ASTM C 581	Immersion in Ca (CO$_3$) at pH=9.5 & 73 ± 3° F	1,000 and 3,000 hours		
Dry heat resistance	ASTM D 3045	140 ± 5° F	1,000 and 3,000 hours		

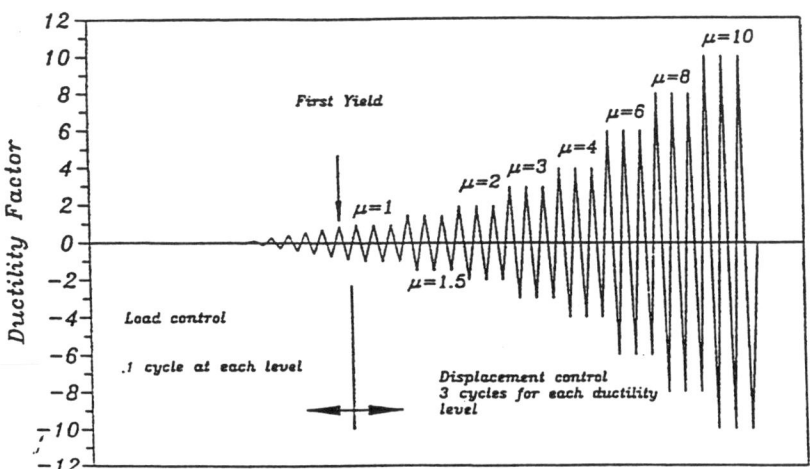

FIGURE 1—TEST SEQUENCE OF IMPOSED DISPLACEMENT

A Design Approach for FRP Composite Structural Shapes

V. SHEKAR, H. THIPPESWAMY and H. V. S. GANGARAO

ABSTRACT

Fiber and Fabric reinforced polymer (FRP) composite materials and structural components are gaining acceptance in civil engineering applications. One application is replacement of conventional concrete bridge decks with FRP structural shapes. This paper presents a simplified design approach for FRP composite structural shapes. The design consists of computing bending and shear rigidities, failure strength, buckling resistance of webs and interaction checks for combined axial, bending and shear effects. Bridge engineers can readily adopt the guidelines provided in this simplified design approach for FRP composite structural shapes without loss of accuracy. As a case study, design of FRP composite H-Deck (SuperDeck^{TM}) is presented herein.

KEY WORDS: Fiber Reinforced Polymer Composites, Classical Lamination Theory, Rovings, Fabrics, Resins, Stiffness, Strength, Stability.

1. INTRODUCTION

Fiber and fabric reinforced polymer (FRP) composite materials and structural shapes are gaining acceptance as viable alternatives to conventional materials in civil infrastructures. These materials are found to have superior performance over conventional materials because of their high strength to weight ratio, excellent fatigue and corrosion resistance. One application is replacement of conventional concrete bridge decks with FRP composite structural shapes (cellular FRP composite deck panels). The flexural rigidity of one such FRP composite deck is found to be about 3.7 times the flexural rigidity of cracked concrete deck (GangaRao et.al 1999).

The structural behavior of FRP composite structural shapes in terms of strength, stiffness, dynamic response, local buckling of web and other factors should be considered in the design. A simplified design approach for FRP composite structural shapes has to be developed so that bridge engineers can readily adopt such an approach with some confidence level. In this paper a simplified design approach is developed based on

---

Vimala Shekar, Graduate Research Assistant, Constructed Facilities Center, West Virginia University, Morgantown, WV 26505.
Hemanth Thippeswamy, Research Assistant Professor, Constructed Facilities Center, West Virginia University, Morgantown, WV 26505.
Hota V. S. GangaRao, Professor of Civil Engineering and Director, Constructed Facilities Center, West Virginia University, Morgantown, WV 26505.

combined experimental and analytical studies conducted on FRP composite structural shapes (I beam, Box-beam, Cellular deck) at the Constructed Facilities Center, West Virginia University (Nagraj, 1994).

2. OBJECTIVES

The objective of this paper is to present a design approach for FRP composite structural shapes. The design procedure presented herein comprises of:
- Computing bending rigidity (EI) and shear rigidity (GA)
- Determining failure strength
- Checking local stability (buckling of web)
- Satisfying the combined effects of axial, bending and shear forces
- Satisfying the deflection limit state

3. DESIGN APPROACH

Two different design philosophies can be adopted in the design of FRP composite structural shapes (Working Stress Design and Load-Resistance Factor Design). However, the proposed approach in this paper is the Working Stress Design methodology because no adequate database is available on short and long term performance of FRP composite bridge decks to establish material resistance factors. In the Working Stress Design approach, safety is covered by attaching factors to ultimate stresses of materials, i.e. allowable stresses, and making sure that those allowable stresses are above the induced stressed obtained from design loads of current building or bridge design code. In addition, safety factors are introduced on ultimate (failure) strains of laminates. Strains in any fiber shall not exceed 20% of the minimum guaranteed long-term laminate strains. Stress level in a laminate shall be also limited based on appropriate failure criterion. First ply failure criterion like Tsai-Hill or Tsai-Wu criterion (Barbero 1998) are used to establish limiting failure stresses. However, it might be necessary to consider failures other than first ply failure in a design by providing greater importance on energy absorption capabilities. Further, the proposed design approach shall incorporate knock-down factors to account for any loss of strength or stiffness during the service-life of FRP composite structural shapes. Some of the knock down factors that should be considered in the design are to account for aging, stress concentrations at re-entrant angles or holes, moisture effects, size effects, sustained stress, etc. For now in the absence of extensive field date this knock-down factor can be taken as 0.4 for FRP composite structural shapes (Karbhari, 1997).

Any design involves computing: (1) bending and shear rigidities; (2) strengths of FRP composite structural shapes; (3) buckling resistance of webs; (4) interaction checks of axial, bending and shear stresses to establish allowable design stresses; and (5) limiting the load induced stresses and deformation within the allowable values. Additional details on the design philosophy can be found from "Philosophy of Design" (GangaRao, 1999) These concepts are integrated and illustrated through a design example in the following section

3.1 Computation of Bending and Shear Rigidities (EI, GA)

Bending and shear rigidities are usually computed by Classical Lamination Theory (Jones, 1975). The in-plane stiffness matrix [A], bending-extension coupling stiffness matrix [B] and bending stiffness matrix [D] (Nagraj, 1994) are computed to predict the bending and shear rigidities of an FRP composite structural shape. The Classical Lamination Theory is modified to simplify the computation of bending and shear rigidities for a FRP composite structural shapes. The modified version known as Approximate Classical Lamination Theory (Nagraj, 1994) is used in this paper. The following are the steps involved in the computation of bending and shear rigidities of a FRP composite structural shapes.

Step1: Compute Material Properties

The material properties include Young's modulus, shear modulus and density of fibers/fabrics and matrix. The properties of fibers and matrix are usually acquired from material suppliers.

Modulus of elasticity of fiber (psi) = E_f
Modulus of elasticity of Matrix (psi) = E_m
Shear modulus of fiber (psi) = G_f
Shear modulus of Matrix (psi) = G_m

From the above properties, Poisson's ratio for fiber a constituent material can be obtained as: (Jones, 1975).

Poisson's ratio of fiber : $\quad v_f = \dfrac{E_f}{2G_f} - 1$

Poisson's ratio of matrix : $\quad v_m = \dfrac{E_m}{2G_m} - 1$

STEP 2: Determine Composite Thickness

Each component (flange and web) is built typically with unidirectional fibers (rovings), randomly oriented fibers (chopped strand mat) and fabrics or a combination of fibers and fabrics. Composite thickness of each ply in the laminate depends on the weight of fibers/fabrics. On an average 40 oz/yd^2 of fabric yields through pultrusion a composite of about 0.05 inch thickness and 3 rovings/inch can result in a composite of about 0.03 inch thickness. Accuracy of thickness depends on manufacturing process, typically given by the manufacturer.

STEP3: Compute Fiber Volume Fraction

Based on the thickness and weight of fiber/fabric of each ply in the composite part, fiber volume fraction is computed as follows:

<u>For Rovings</u>

$$V_f = \frac{n\pi D^2}{4bt} \qquad (EQ.1)$$

Where,

n = Number of bundles
b = Width of laminate (in)
t = Thickness of composite layer (in)
D = Diameter of fiber = $\sqrt{\dfrac{1}{\rho_f Y 9\pi}}$
ρ_f = density of fiber in lb/in^3
Y = yield (a number in yards which weighs 1 lb)

<u>For CSM (Continuous Strand Mat) and Fabric</u>

$$V_f = \frac{W_f}{\rho_f L_v} \qquad (EQ.2)$$

Where,

W_f = Weight of CSM/Fabric (lbs)
L_v = Volume of 1' x 1' composite laminae (in^3)
ρ_f = Density of CSM or fabric (lbs/in^3)

STEP 4: Evaluate Laminate Properties

The stiffness properties of a laminate are computed by Rule-of -Mixture (Jones, 1975)

<u>Properties for Fabric and Rovings</u>

Longitudinal Modulus (psi) $\qquad F_{11} = E_j V_j + E_m (1 - V_f) \qquad$ (EQ.3)

| Transverse Modulus (psi) | $E_{22} = \dfrac{E_f E_m}{E_f V_m + E_m V_f}$ | (EQ.4) |

(inverse of Rule-of-Mixture)

| In-plane Shear Modulus (psi) | $G_{12} = \dfrac{G_f G_m}{G_f V_m + G_m V_f}$ | (EQ.5) |

| In-plane Poisson's Ratio | $v_{12} = v_f V_f + v_m (1 - V_m)$ | |

$$v_{21} = \dfrac{v_{12} E_{22}}{E_{11}} \quad \text{(EQ.6)}$$

For Continuous Strand Mat

Elastic Modulus (psi)	$E_{ran} = \dfrac{3}{8} E_{11} + \dfrac{5}{8} E_{22}$	(EQ.7)
Shear Modulus (psi)	$G_{ran} = \dfrac{1}{8} E_{11} + \dfrac{1}{4} E_{22}$	(EQ.8)
Poisson's Ratio (psi)	$v_{ran} = \left(\dfrac{E_{ran}}{2 G_{ran}}\right) - 1$	(EQ.9)

STEP 5: Compute E_x

Modulus of laminate in the direction of bending (along the fiber direction) is determined in an approximate manner by (Nagraj 1994)

$$E_x \approx E_{11} \cos^4(\theta) \quad \text{(EQ.10)}$$

Where 'θ' is the angle of fiber orientation with respect to bending direction. The global and material coordinate systems are represented in Figure 1.

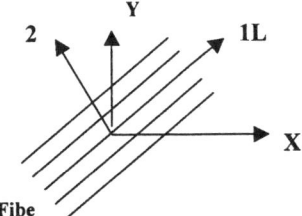

Note: X, Y are Global axis
1, 2 are Local axis

Figure 1: Local and Global Coordinate Systems

STEP 6: Compute In-plane Stiffness [A]

$$A_f = A_w = b\sum_{k=1}^{N}(E_x)_k t_k \qquad (EQ.11)$$

Where,

$(E_x)_k = E_x$ in k^{th} layer, where 'x' corresponds to global axis
t_k = thickness of the kth layer (in)
b = width of laminate (in)

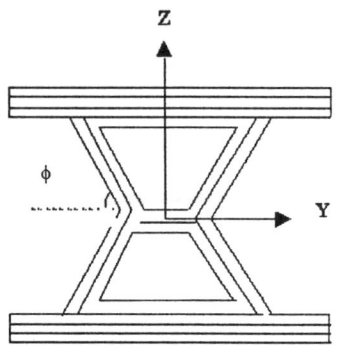

Figure 2 Cross-Section of FRP Composite deck (SuperDeck^{TM}) with 'k' layers

STEP 7: Compute Extensional-Bending Coupling Stiffness [B]

$$B \approx b\sum_{k=1}^{N}(E_x)_k t_k Z_k \qquad (EQ.12)$$

Where 'Z_k' is distance of mid-surface of kth lamina from the centroid of the section.

STEP 8: Compute Flange and Web Bending Stiffness

For flange (Nagraj, 1994):

$$D_f \approx b\sum_{k=1}^{N}(E_x)_k \left[t_k Z_k^2 + \frac{t_k^3}{12} \right] \qquad (EQ.13)$$

For web (Lopez, 1995):

$$D_w \approx b \sum_{k=1}^{N} (E_x)_k \left[\left(\frac{t_k^3}{12} + t_k Z_k^2 \right) \cos^2(\phi) + \left(\frac{b^2 t_k}{12} \right) \sin^2(\phi) \right] \quad \text{(EQ.14)}$$

Where,
'ϕ' is angle of the component with respect to the horizontal;
'f' refers to flange and 'w' refers to web

STEP 9: Compute Global Bending Stiffeness (EI) in X direction

$$EI \approx \sum_{f=1}^{n} [D_f + A_f e_f^2] + \sum_{w=1}^{m} [D_w + A_w e_w^2] \quad \text{(EQ.15)}$$

Where,
n = number of flanges
m = number of webs
e_f = eccentricity of a flange or web from the mid-surface of component

STEP 10: Compute Global Shear Stiffness (GA) in XY plane

$$GA = d \sum_{k=1}^{N} (G_x)_k t_k \quad \text{(EQ.16)}$$

Where
$(G_x) \approx E_{11} \sin^2\theta \cos^2\theta + G_{12}(\sin^2\theta - \cos^2\theta)^2$

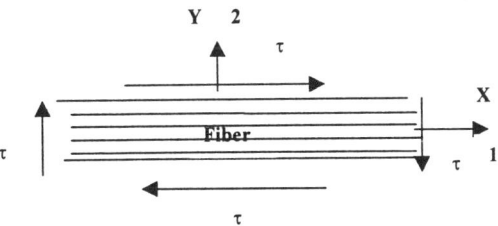

Figure 3 Determination of G_{12}

Note: X and Y refer to global axis and 1 and 2 refer to local axis

$(G_x)_k$ = shear in k^{th} layer (psi)
t_k = thickness of kth layer (in)
d = depth of the laminate (in)

The global shear stiffness formulae is accurate upto fiber orientation of 45^0 and results in a conservative value for shear stiffness. The shear stiffness in reality has to be higher than the approximated formulae as E_{22} effect (Transverse Modulus) is neglected. We suggest using Classical Lamination Theory (CLT) to compute accurate shear stiffness. The results are conservative only for shear stiffness and not for bending stiffness.

3.2 Computation of Failure Strength

STEP 11: Failure Strength

The failure strength is computed as per working stress design approach. From the experimental test results (Vedam, 1997) the ultimate bending strain for FRP composite coupons was found to be 18000 ~ 20000 microstrain. From the working stress design approach, only 20% of ultimate strain is considered in evaluating the strength of FRP composite deck. The strength, should be further reduced by a knock-down factor of 0.4 (Karbhari, 1997).

Failure strength $\sigma = 0.4(0.2\varepsilon \times E)$ (EQ.17)

Where,
ε = ultimate strain (microstrains)
E = Young's modulus (psi)
I = Moment of inertia about axis of bending (in^4)

The failure strength of FRP composite structural shapes should also be limited by failure criterion (Tsai-Hill/Tsai-Wu). The failure criterion as per Tsai-Hill/ Tsai-Wu is given in (Barbero, 1998). However, based on our extensive experimental data, 20000 microstrain is taken as the limiting failure stain in bending while in shear the failure strain is taken as 10000 microstrain,

3.3 Check for Local Stability

STEP 12: Local Stability

The FRP composite structural shapes on occasions can fail prematurely. This is attributed to poor fiber architecture and manufacturing process, which eventually lead to local failure such as buckling of web or flange, delamination at the web-flange junction, etc. Hence, the web should have an adequate thickness to overcome local buckling. Inadequate web thickness will also lead to premature failure coupled with bending in web or flange. The following checks should be satisfied to ensure local stability.

<u>Check for axial stresses</u>
Induced axial stress < Critical axial stress

Induced axial stress $\sigma_{ia} = \dfrac{P}{A}$ (EQ.18)

Where

P = compressive load on the web (lbs)
A = cross-section of web (in²)

Critical axial stress $\sigma_{ca} = \dfrac{k\pi E}{12(1-v^2)}\left(\dfrac{t_w}{d_w}\right)^2$ (EQ.19)

Where,
k = coefficient depending on end conditions
E = modulus of elasticity (psi)
v = Poisson's ratio
t_w = thickness of web (in)
d_w = depth of web (in)

Check for bending stresses
Induced bending stress < Critical bending stress

Induced bending stress $\sigma_{ib} = \dfrac{Mc}{I}$ (EQ.20)

Where,
M = bending moment acting on the web (lbs-in)
c = $t_w/2$
I = moment of inertia = $bt_w^3/12$

Critical bending stress $\sigma_{cb} = \varepsilon_w E$ (EQ.21)

ε_w = bending strain ranging from 18000 ~ 20000 microstrains

Check for shear stresses
Induced shear stress < Critical shear stress

Induced shear stress $\tau_i = \dfrac{1.5V}{bt_w}$ (EQ.22)

Where,
V = shear load (lbs)

Critical shear stress $\tau_c = \varepsilon_s G$ (EQ.23)

Where,
G = shear modulus (psi)
ε_s = shear strain

The ultimate shear strain is about 20000 microstrain (Wen 1999), but the strain does not increase linearly, i.e., the stress versus shear stress strain curve for a GFRP composite sample is nonlinear. Hence, the average shear strain is taken as 10000 mircrostain.

3.4 Check for Interaction Equations

STEP 13: Interaction Equations

The interaction equation for axial, bending and shear is well established for steel structures. Since we do not have enough experimental data for the combined effect of axial, bending and shear on composite structures, the interaction equation for composites in the current design is considered to be same as that of steel.

$$\left(\frac{\sigma_{ia}}{\sigma_{ca}}\right) + \left(\frac{\sigma_{ib}}{\sigma_{ca}}\right)^2 + \left(\frac{\tau_i}{\tau_c}\right)^2 \leq 1 \qquad (EQ.24)$$

Note: EQ.24 is a modification of the interaction equation given in ASCE Structural Plastics Design Manual for shear and bending stresses.

3.5 Check for Deflection

STEP 14: Deflection

Deflections are significant in FRP composites structural shapes because of their low stiffness compared to structural shapes made of conventional materials. The deflection limit is given as follows:
Deflection of bridge superstructure under live loads < span/800
Deflection for bridge decks under live loads < span/360

4. DESIGN COMPUTATIONS FOR SUPERDECK™

Laurel Lick and Wickwire Run are the two bridges, which were constructed with a modular FRP composite deck (SuperDeck™). The SuperDeck™ comprised of double trapezoid and hexagonal components. The cross-section of the SuperDeck™ is shown in Fig 1. The thickness of flange, wing, core and web in double trapezoid component (FT1, FT2, FT3, FT4) and flange and web in hexagonal component (FH1) are shown in Table 1. E-glass fiber and Vinyl ester resin are the main constituents of the SuperDeck™. The components are mainly built with rovings, chopped strand mat, and fabric (CFM, 90^0, $\pm 45^0$).

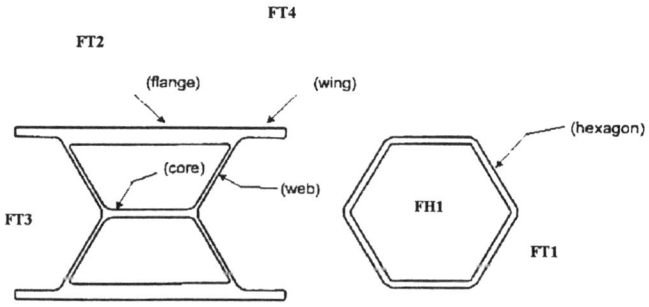

Figure 1 Cross section of SuperDeck™

A Design Approach for FRP Composite Structural Shapes

Table 1 Details of SuperDeck™ Components

Component	Section	Thickness (in)
Double-trapezoid	FT1	0.75
	FT2	0.438
	FT3	0.375
	FT4	0.228
Hexagon	FH1	0.3125

Bending and Shear Rigidity (Step 1 through Step 10)

The cross-section is divided into individual components for ease of computation. The stiffness of each component is determined and is then added using the principle of parallel axis theorem. The theoretical bending and shear rigidities are computed as per step1 through step 10. Comparison of theoretical values with experimental values are given in Table 2.

Table 2 Comparison of experimental and analytical component rigidities

Components	Experimental		Analytical	
	Bending rigidity (EI) lbs-in^2	Shear rigidity (GA) lbs	Bending rigidity (EI) lbs-in^2	Shear rigidity (GA) lbs
Double-trapezoid	6.72 x 10^8	3.99 x 10^6	6.99 x 10^8	1.67 x 10^6
Hexagon	1.40 x 10^8	1.17 x 10^6	1.45 x 10^8	2.13 x 10^6

Failure Strength (Step 11)

Strength of FRP bridge deck component (double-trapezoid) $\sigma = 0.4(0.2\varepsilon \times E)$
ε = assuming 20000 microstrain
I = Moment of Inertia

$$I \approx 2\left[\frac{12 \times 0.75^3}{12} + (12 \times 0.75) \times (3.625)^2\right] + \left[\frac{4 \times 0.375^3}{12}\right] +$$
$$4\left[\frac{4 \times 0.228^3}{12}\cos^2 60 + \frac{4 \times 0.228^3}{12}\sin^2 60 + (4 \times 0.228) \times (2)^2\right] = 250 in^4$$

Modulus of elasticity $E = \dfrac{6.99 \times 10^8}{250} = 2.8 \times 10^6$ psi

Failure strength $\sigma = 0.4(0.2 \times 20000 \times 10^{-6} \times 2.8 \times 10^6) = 4480 psi$

Check for local Stability (Step 12)
The double-trapezoid component is checked for local web buckling. The induced stress due to axial, bending and shear should be less than the allowable stress in the FRP wall as mentioned in Section 3.3. The failure load of 84" long double-trapezoidal component under 3-point bending with a patch (10" x 20") load was found to be 30 kips (Vedam, 1997).

As per EQ.18, EQ.20 and EQ.22 under 30 kips load, induced stresses were computed.
Induced axial stress = 2254 psi
Induced bending stress = 54557 psi and
Induced shear stress = 2059 psi
After establishing critical axial (13607 psi), bending (58600 psi) and shear stress (9500) as per EQ.19, EQ.21 and EQ.23, the following interaction checks was given:

Check for Interaction Equations (Step 13)

$$\left(\frac{2254}{13607}\right) + \left(\frac{54557}{58600}\right)^2 + \left(\frac{2059}{9500}\right)^2 = 1.07 > 1$$

Since the combined effect of axial, bending and shear effect is more than 1, the web thickness of the double-trapezoid shape has to be revised.

Check for Deflection (Step 14)
Experimental deflection of a composite deck specimen of 36" x 8" (double-trapezoid) under static load for a span length of 9 ft. at a load of 35 kips was 0.375 inches.
Accounting for effective deck width (EW) = 48" in a FRP deck and assuming continuous over stringers (providing a continuity factor CF = 0.75),
Effective deflection = (0.375)(CF)/(EW) = (0.375)(0.75)/(48/36) = 0.211 inches. Deflection can be further reduced by about 20% due to plate action; however such reduction is not shown here.
Allowable deflection = (9x12)/360= 0.3 inches.
The deflection is well within the allowable limits.

5. CONCLUSIONS

- A simple design procedure has been developed for FRP structural shapes and its accuracy has been validated through experimental data.
- The Approximate Classical Lamination Theory (ACLT) used to compute bending stiffness and shear rigidity of FRP composite structural shapes (e.g. double-trapezoid), are in good correlation with experimental results (Table 2).
- Inadequate web thickness in FRP structural shapes can lead to local failure.
- The deflection is within allowable limits.

REFERENCES

Ever J. Barbero (1998), " Introduction to Composite Materials Design", Taylor & Francis Inc.

GangaRao, H.V.S et al; (1999), " Development of Glass Fiber Reinforced Polymer Composite Bridge Deck", SAMPE Journal v35 n4

Jones, R.M. (1975), "Mechanics of Composite Materials", Hemisphere Publishing Co.

Karbhari, V.M. and Seible F; (1997), International Symposium on Non-metallic Reinforcement for Concrete Structures, Sapparo, Japan, pp 191-198.

Nagraj, V. (1994), "Static and Fatigue Response of Pultruded FRP Beams Without and With Splice Connections," MS Thesis, Department of Civil and Environmental Engineering, West Virginia University, Morgantown, WV 26506, USA.

"Structural Plastics Design Manual" (1984) ASCE

Vedam, V.R. (1997), " Characterization of Composite Material Bridge," MS Thesis, Department of Civil and Environmental Engineering, West Virginia University, Morgantown, WV 26506, USA.

Repairing and Strengthening Reinforced Concrete Columns Using Ferrocement Laminates

E. H. FAHMY, Y. B. I. SHAHEEN and Y. S. KORANY

ABSTRACT

This paper presents a proposed method for repairing reinforced concrete columns using ferrocement as a viable economic alternative to the highly expensive conventional jacketing methods by reinforced concrete or steel jackets. The results of both experimental and analytical investigations to examine the effectiveness of the proposed method are reported and discussed including strength, deformation, ductility ratio, and energy absorption characteristics of the repaired specimens.

Twenty-four reinforced concrete column specimens were tested under concentric compression load. Each specimen was first loaded till failure or up to either 67% or 85% of the ultimate load of the control specimens. After unloading, the damaged column specimens were repaired by a complete jacket form of 10mm thick ferrocement around the four sides of the specimen. Three different types of reinforcing steel meshes were used, welded wire mesh and expanded metal meshes type X8 and EX156. The investigated parameters included the pre-loading level, type of reinforcing mesh, mesh opening direction, and volume fraction of mesh reinforcement.

The finite element technique was used to model the repaired columns in order to investigate the essential parameters. The 3-dimensional feature available in the analysis package ANSYS 5.4 was used to model all test specimens. The model accommodates the material non-linearity, cracking and crushing of concrete and yielding of the steel. The analytical results compared well with the experimental ones, which served as verification for the analytical model.

The analytical and experimental results of the repaired columns demonstrated that irrespective of the pre-loading level or the repair scheme, better load carrying capacity for all test specimens could be achieved compared to their original behavior.

Ezzat H. Fahmy, Professor, Engineering Department, The American University, Cairo, Egypt
Yousry B.I. Shaheen, Associate Professor, Faculty of Engineering, Menoufia University, Shebin El-Kom, Egypt.

Yasser S. Korany, Teaching & Research Assistant, Civil Engineering Department, University of Manitoba, Winnipeg, Canada. Former Teaching & Research Assistant, Engineering Department, The American University, Cairo, Egypt

Under short term loading conditions, all repaired specimens restored more than their original ultimate strengths. It was found that the level of damage sustained prior to repairing affects the ultimate load of the repaired column specimens.

INTRODUCTION

Many factors contribute to the deterioration of reinforced concrete structures and affect their safety. Some of these are cracking due to impact and dynamic loading, creep, thermal cycling, inadequate design or faulty construction. Many investigations have been undertaken in strengthening and repair of slabs and beams but only little research work is available for strengthening and repairing of columns. Because of its ease of application and low cost, especially in developing countries, ferrocement was used as a repair material for reinforced concrete elements as an alternative to the other highly expensive repair materials.

Mansur and Paramasivam [1] carried out an experimental investigation on ferrocement box- section short-columns with and without concrete infills under axial and eccentric compression. The major parameters were types, arrangements, and volume fraction of reinforcement. Test results indicated that a ferrocement box-section could be used as a structural column. Welded wire mesh has been found to perform better than an equivalent amount of woven mesh.

Kaushik et al [2] carried out an investigation on the ferrocement encased concrete columns. They have investigated short circular as well as square columns with unreinforced and reinforced cores. It was seen that the ferrocement encasement increases the strength and ductility of the columns for both axial and eccentric loading conditions.

Another interesting research work was done by Ahmed et al. [3] to investigate the possibility of using ferrocement as a retrofit material for masonry columns. Uniaxial compression tests were performed on three uncoated brick columns, six coated brick columns with 25 mm plaster and another six columns coated with 25 mm thick layer of ferrocement. The study demonstrated that the use of ferrocement coating strengthens brick columns significantly and improves their cracking resistance.

Nedwell et al. [4] conducted a preliminary investigation into the repair of short square columns using ferrocement. A short program was undertaken to provide some information regarding the effect of ferrocement repair on short columns subject to axial loading. It was found that the use of ferrocement retrofit coating increases the apparent stiffness of the columns and significantly improves the ultimate load carrying capacity.

The work presented in this paper is the third phase of investigating the use ferrocement in repairing and retrofitting of reinforced concrete elements. The first and second phases included repairing reinforced concrete slabs and beams (Fahmy et al. [5] & [6]. This phase focused on repairing reinforced concrete short columns since the majority of columns in buildings are short ones.

EXPERIMENTAL PROGRAM

In addition to the three control specimens, twenty-one column specimens were cast and divided into groups of three specimens. All test specimens had cross-sectional dimensions, before repair, of 10cmX10cm and 100cm in length. The concrete mix was designed to obtain an ultimate compressive strength at 28-days age of 30 MPa. The mix proportions by weight were (1.0:0.4: 0.8) for cement: fine aggregate: coarse aggregate, and the water cement ratio was 0.5. The concrete slump was found to be 46mm while the air entrainment was found to be 2.3% with a density of 2407 kg/m^3. Four mild steel bars of 6mm diameter having yield strength of 2400 kg/cm^2 were used to reinforce the test specimens in addition to six stirrups 2.7mm diameter at 20cm intervals. To provide more safety against undesirable column head failure during testing, 2 additional stirrups were used at each column head. All specimens were soaked with water twice a day for 14 days followed by air curing for another 14 days before being pre-loaded.

All specimens were tested under central uniaxial compression in a hydraulic universal testing machine with a sphere loading head. After placing the specimen in the testing machine, vertical alignment was adjusted to eliminate any eccentricity. Loads were recorded along with column deformation through the attached data acquisition system. The test setup for a typical column specimen is shown in figure(1). After unloading, each specimen was repaired by thin ferrocement jacket consisting of 10.0 mm mortar reinforced with closely spaced steel mesh as shown in figure(2). The sand-cement mortar of ferrocement consisted of sand, ordinary Portland cement, and silica fume with sand cement ratio of 2.0. The water/cementitous materials ratio used was 0.35. Based on the results of previous research, Korany [7], 15% of cement was replaced by silica fume to reach as high strength mortar as possible. Superplasticizer with ratio of 1.2% by weight of cement/silica fume was used to improve workability.

The test specimens were divided into eight groups according to the pre-loading level reached prior to repairing and the type of steel mesh used. Details of test groups are given in Table I. In addition to the welded wire mesh, two different types of expanded metal lath were used to reinforce the ferrocement laminates. The first expanded type is X8 with opening size 9.5X31mm and weight of 3.54 kg/m^2 while the second type is EX156 with opening size 19X63mm and weight of 2.73 kg/m^2. The welded wire mesh has a grid of 25X25mm with wire diameter of 2.7mm and weight of 3.57 kg/m^2. The yield stresses of the welded wire mesh and expanded metal lath types X8 and EX156 were experimentally determined and found to be 2400 kg/cm^2, 1700 kg/cm^2, and 1000 kg/cm^2, respectively. The reinforcing mesh of the ferrocement layer was secured to the specimen by 5 dowels at each side, previously imbedded in specimens during casting, having a diameter of 2.7mm to provide composite action between column specimens and the ferrocement jacket. In practical applications this could be achieved by means of screw bent nails. The mesh used was first pre-folded providing an overlap of 5cm at the ends. Then the mesh was wrapped around the column specimen and the ends were secured together using double thin steel wires commonly used in tying reinforcing bars. The sand-cement matrix was hand-applied to form a 10.0mm thick ferrocement layer all around the specimen.

The three control column specimens C_C were loaded to failure to determine the ultimate load (P_u) and load-deformation relationship. These three specimens were then repaired and designated C_1.

For the repaired columns, each specimen was loaded up to 67%, 85% or 100% of the ultimate load of the control column as shown in Table I; then it was unloaded and the ferrocement layer was applied.

Figure 1. Test Set-up and Typical Specimen

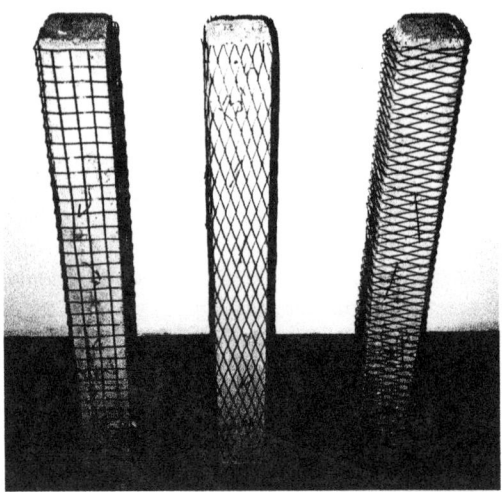

Figure 2. Different Steel Mesh Types

Table I. DETAILS OF TEST GROUPS

Group Designation	Pre-loading level	Repair Type	Mesh Type	No of Meshes	V_f*
C_1	100%	complete jacket	Exp-X8 L	1	0.0453
C_2	85%	complete jacket	Exp-X8 L	1	0.0453
C_3	85%	complete jacket	Exp-156 L	1	0.0350
C_4	85%	complete jacket	Exp-156 T	1	0.0350
C_5	85%	complete jacket	Welded	1	0.0458
C_6	67%	complete jacket	Exp-X8 L	1	0.0453
C_7	67%	complete jacket	Exp-156 L	1	0.0350
C_8	67%	complete jacket	Welded	1	0.0458

* *Volume of reinforcement/ Volume of composite (Ferrocement + reinforcement).*

ANALYTICAL MODEL

A finite element model was developed to simulate the repaired columns to be used to extend the parametric study. An educational version 5.4 of ANSYS [8], a multi-purpose finite element package, was used in the present analytical investigation. Due to symmetry, only a quarter of the specimen was modeled. Displacement was applied at the nodes of the top surface of the model and the summation of the resulting reactions at the bottom surface was calculated. A full transient analysis was performed in 3 load steps: loading of the original column up to the desired load level, unloading the column, then reloading the column model after activating the repair layer. The element "Birth" and "Death" feature of the ANSYS program was used to add the desired repair layer to the original models. The repair layer elements were deactivated "Killed" during the first loading and unloading steps, subsequently they were activated "Born" in the last load step.

The three dimensional element SOLID65, available in the ANSYS elements library, was used to model both concrete and ferrocement layer. SOLID65 is defined by 8-nodes having three degrees of freedom at each node: translations in the x, y, and z directions (ux, uy, and uz). The element is capable of cracking in tension and crushing in comparison. The rebar reinforcement feature of this element was used to model the mesh reinforcement of ferrocement while the 2-nodded element LINK8 was used to model the discrete steel bars and stirrups used to reinforce the original column. LINK8 is a uniaxial tension-compression element with three degrees of freedom: ux, uy, and uz at each node.

The first crack strength for the elements containing the mesh reinforcement of the ferrocement layer was estimated from the theoretical model proposed by Paramasivam and Nathan [9] as:

$$f_{cr} = f_s(A_s/A_c)^n + f_t \qquad (1)$$

Where (f_{cr} and f_t) are the respective cracking strength of the composite and the mortar matrix, (f_s) is the steel yield stress or the proof stress at 0.01% strain, and (A_c and A_s) are the cross-sectional area of the composite and steel while (n) is a constant proposed to be 1.3. The tensile strength of the ferrocement mortar was determined experimentally and found to be 5.4 Mpa.

The three-dimensional point to point contact element CONTAC52 was chosen to represent the contact between the outer surface of the concrete column and the ferrocement layer. The element is capable of supporting only compression in the direction normal to the contact surfaces and shear in the tangential direction. A gap of 0.1mm was specified between the two nodes of the element, which may have one of three conditions: closed and stuck, closed and sliding, or open.

RESULTS AND DISCUSSION

Results of the Experimental Program

The ultimate load (P_u), maximum deformation, ductility ratio and energy absorption for all test specimens are listed in Table II while the average results and percentage gain for each column set were calculated and summarized in Table III.

Ductility ratio was defined, in this paper, as the ratio of the maximum deformation at ultimate load to that at the onset of yielding while energy absorption was defined as the area under the load-deformation curve up to failure. It should be noted here that the control specimens, which were pre-loaded to failure, were also repaired and tested for comparison purposes. The repaired control specimens will be referred to as column group C_1. Figure(3) shows the average load-deformation curve for the control group after repairing, group C_1. Figure(4) compares the average load-deformation curves of groups C_2, C_3, C_4 and C_5 pre-loaded to 85% of the ultimate load. While Figure(5) compares the average load-deformation curves of groups C_6, C_7 and C_8 pre-loaded to 67%.

Table II. EXPERIMENTAL RESULTS OF COLUMN SPECIMENS

Group Designation	Column No	P_u (kN)	Δ_{max} (mm)	Ductility Ratio	Energy Absorption (N.m)
Cc	Cc-1	145.6	4.80	1.02	306
	Cc-2	149.5	4.90	1.08	358
	Cc-3	155.5	4.70	1.05	389
C_1	C_{1-1}	231.5	7.57	1.27	618
	C_{1-2}	205.1	7.31	1.30	607
	C_{1-3}	224.9	7.31	1.42	613
C_2	C_{2-1}	220.2	7.47	1.27	702
	C_{2-2}	229.0	7.38	1.24	676
	C_{2-3}	231.8	7.56	1.20	669
C_3	C_{3-1}	212.0	8.51	1.25	934
	C_{3-2}	216.0	8.57	1.29	981
	C_{3-3}	214.3	8.71	1.33	963
C_4	C_{4-1}	198.0	11.17	1.25	933
	C_{4-2}	191.0	11.23	1.27	937
	C_{4-3}	194.8	11.50	1.25	941
C_5	C_{5-1}	270.0	6.23	1.20	683
	C_{5-2}	272.8	6.35	1.16	670
	C_{5-3}	271.4	6.32	1.17	638
C_6	C_{6-1}	249.0	4.72	1.14	733
	C_{6-2}	244.3	4.65	1.17	691
	C_{6-3}	254.6	4.73	1.13	692
C_7	C_{7-1}	220.0	7.42	1.20	1008
	C_{7-2}	222.0	7.61	1.18	978
	C_{7-3}	221.3	7.47	1.22	981
C_8	C_{8-1}	295.0	7.37	1.10	680
	C_{8-2}	280.0	4.40	1.13	656
	C_{8-3}	279.1	4.14	1.16	737

Table III. AVERAGE EXPERIMENTAL RESULTS OF COLUMN GROUPS

Group Designation	Ultimate Load, P_u (kN)	% Increase in P_u	Ductility Ratio	% Increase in Ductility	Energy Absorption (N.m)	% Increase in Energy Absorption
Cc	150.2	-	1.05	-	318.0	-
C1	220.5	46.8	1.33	26.6	612.8	92.7
C2	227.0	51.1	1.23	17.1	682.4	114.6
C3	214.1	42.5	1.29	22.9	962.6	202.7
C4	194.6	29.6	1.26	20.0	937.1	194.7
C5	271.4	80.7	1.18	12.4	663.7	108.7
C6	249.3	66.0	1.15	9.5	705.6	121.9
C7	221.1	47.2	1.20	14.3	989.0	211.0
C8	284.7	89.5	1.13	7.6	691.0	117.3

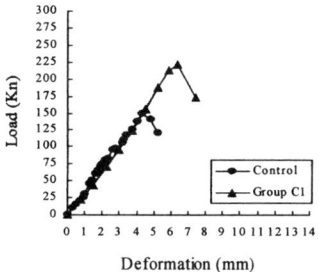

Figure 3. Load-Deformation Curve of the "Repaired" Control Group

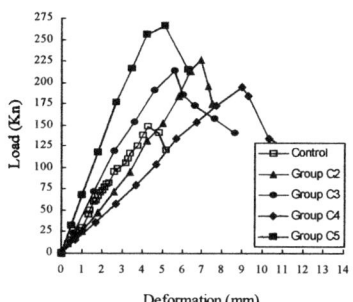

Figure 4. Load-Deformation Curves for Column Groups Pre-loaded to 85% of P_u

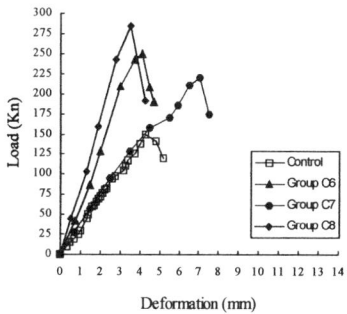

Figure 5. Load-Deformation Curves for Column Groups Pre-loaded to 67% of P_u

Based on the test results, the increase in the ultimate loads after repairing is quite obvious as shown in Table II. However, almost all test specimens showed premature failure due to the platen effect of the testing machine on the column head. Better results could have been reached if the columns' heads were further strengthened or if special end plates were used, Nedwell et al [4]. All repaired column specimens showed similar load-deformation behavior with roughly linear response up to yielding. The results presented in Table III showed increase in the ductility ratio of the repaired columns. Irrespective of the repair scheme applied, all test specimens showed large deformation at ultimate loading which is an indication of high ductility and energy absorption.

Raising the pre-loading level from 67% to 85% and up to 100% resulted in lower gain in P_u from 66% to 51.1% and 46.8% for column groups C_6, C_2, C_1, respectively, having the same steel type and volume fraction (V_f). The main reason for this behavior is that increasing the pre-loading level increases the damage of the column specimens before repairing. Similar behavior was observed for column groups repaired using welded wire mesh; column groups C_8 and C_5 with the gain in P_u reduced from 89.5% to 80.7%. It was noticed for column groups pre-loaded to failure (C_1), and to 85% of the ultimate load (C_2) that the deformation values were higher than those for column group pre-loaded to 67% of the ultimate load (C_6). Also, it is interesting to note the high gain in ductility for group C_1, 26.6%, over the gain of groups C_2 and C_6, 17.1% and 9.5%; respectively. High pre-loading levels resulted in lower gain percentage in energy absorption. This can be seen from the comparison between the results of column groups C_1, C_2 and C_6. At pre-loading level of 67% of the ultimate load, the increase in energy absorption was 121.9% and at pre-loading level 85% the increase was 114.6% while it was 92.7% at pre-loading level 100%.

The test results revealed that raising the reinforcement ratio of the ferrocement layer from 0.035 for expanded EX156 to 0.0453 for expanded X8 results in significant increase in the load carrying capacity. This is clearly seen from comparing the results of column groups C_3 and C_7 with those of column groups C_2 and C_6; respectively. This increase in P_u is directly attributed to the larger V_f that provided the ferrocement jacket with higher strength and provided more confinement to the original column core. Raising the reinforcement ratio of the ferrocement layer from 0.035 to 0.0453 resulted in lower gain in the ductility ratio. This is clearly seen from the comparison between the gain in the ductility ratio of column groups C_3 and C_6 with those of C_2 and C_7. This decrease in ductility was expected since ductility decreases with the increase in the steel ratio. From the experimental results given in Table III, it was also found that raising the volume fraction from 0.035 for column groups C_3 and $C7$ to 0.0453 for column groups C_2 and C_6, resulted in lower gain in energy absorption, from 202.7% and 211% to 114.6% and 121.9%, respectively. This may be attributed to the decrease in ductility accompanying the increase in volume fraction.

A comparison of the results of column groups C_2 and C_6 with those of C_5 and C_8 showed that using welded wire mesh seems to be more efficient than the expanded metal lath since both types have almost the same V_f. The gain in P_u for C_2 and C_6 was found to be 51.1% and 66% while for C_5 and C_8 the gain was 80.7% and 89.5%; respectively. This may be attributed to the better confinement effect of the welded wire mesh than that of the expanded metal lath. The average test results given in Table III showed that the increase in ductility is higher for the expanded metal than that for the welded wire mesh. This is obvious from a comparison of the results of group C_2 (17.1%) with those of group C_5 (12.4%) having almost the same V_f and pre-loaded to 85%. The same trend was noticed when comparing the results of group C_6 (9.5%) with those of group C_8 (7.6%) pre-loaded to 67%. This may be due to the diamond configuration of the expanded metal lath which accepts more deformation and absorbs more energy. In a similar trend to ductility ratio, it was found that the expanded metal exhibits more gain in energy absorption than the welded wire mesh. The gain in energy absorption for groups repaired using expanded metal was found to be 114.6% and 121.9%, C_2 and C_6, while it was 108.7% and 117.3% for those repaired using welded wire mesh, C_5 and C_8.

Also the test results showed that the orientation of the steel mesh opening has a significant effect on the gain in P_u. The gain in P_u for column group C_3 repaired using expanded EX156 in the longitudinal direction was found to be 42.7% while for group C_4 repaired with the same type of mesh but in the transverse direction was found to be only 29.7%. It is interesting to note that at early loading history, larger initial cracks were developed for group C_4. Because of the limited number of available specimens for comparison, further research work is needed before being able to judge the effect of the mesh opening orientation of the expanded metal lath on the strength and performance of the repaired columns. Although the orientation of the reinforcement has significant effect in the gain in P_u, it seems that it has no effect on ductility. This can be realized from a comparison of the results of group C_3 with those of group C_4. The explanation could be that both groups had similar steel ratios which is the main factor influencing ductility. For the same volume fraction, the test results revealed that the orientation of the reinforcement has a minor effect on the energy absorption. The gain in energy absorption for column group C_3 repaired using expanded EX156 in the longitudinal direction was found to be 202.7% which is very close to the gain percentage of group C_4 repaired by the same type but in the transverse direction, 194.7%. This result is in agreement with the gain in ductility.

Results of the Analytical Study

The described analytical model was used to analyze all test specimens. The analytical results of all column groups tested experimentally are listed in Table IV while a comparison between the predicted and experimental results is given in Table V and figures (6), (7), and (8).

Table IV. PREDICTED RESULTS OF COLUMN MODELS

Group Designation	P_u (kN)	% Increase in P_u	Ductility Ratio	% Increase in Ductility	Energy Absorption (N.m)	% Increase in Energy Absorption
Cc	158.2	-	1.02	-	259.5	-
C1	224.5	41.90	1.20	17.65	512.9	97.7
C2	244.0	54.24	1.13	10.80	675.6	160.3
C3	225.0	42.23	1.17	14.71	682.0	162.9
C4	206.8	30.70	1.20	17.65	649.8	150.4
C5	275.0	73.80	1.30	27.50	496.5	91.3
C6	265.0	67.51	1.24	21.60	628.6	142.2
C7	225.0	42.23	1.14	11.76	747.6	188.1
C8	291.0	83.94	1.26	23.53	609.5	134.8

Table V. COMPARISON BETWEEN EXPERIMENTAL AND PREDICTED RESULTS

Group	P_u (kN)			Ductility Ratio			Energy Absorption (N.m)		
	Exp*	Pred**	$\frac{Pred}{Exp}$	Exp.	Pred.	$\frac{Pred}{Exp}$	Exp.	Pred.	$\frac{Pred}{Exp}$
Cc	150.2	158.2	1.053	1.05	1.02	0.97	270.3	259.5	0.96
C1	220.5	224.5	1.018	1.33	1.20	0.90	551.5	512.9	0.93
C2	227.0	244.0	1.075	1.23	1.13	0.92	614.2	675.6	1.10
C3	214.1	225.0	1.051	1.29	1.17	0.91	749.7	682.0	0.91
C4	194.6	206.8	1.063	1.26	1.20	0.95	722.0	649.8	0.90
C5	271.4	275.0	1.013	1.18	1.30	1.15	564.2	496.5	0.88
C6	249.3	265.0	1.063	1.15	1.24	1.08	635.0	628.6	0.99
C7	221.1	225.0	1.017	1.20	1.14	0.95	890.1	747.6	0.84
C8	284.7	291.0	1.022	1.13	1.26	1.12	621.9	609.5	0.98

* Exp. = Experimental
** Pred. = Predicted

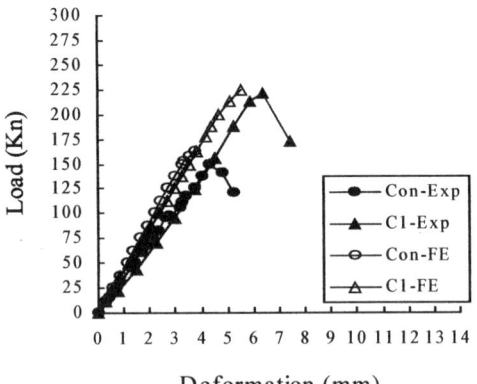

Figure 6. Comparison between Load Deformation Curves of the Control Group before and after Repairing

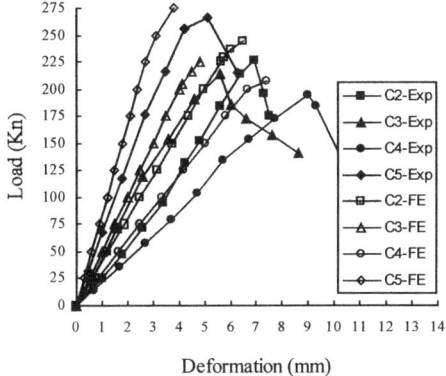

Figure 7. Comparison between Load-Deformation Curves of Column Groups Pre-loaded to 85% of The Ultimate Load

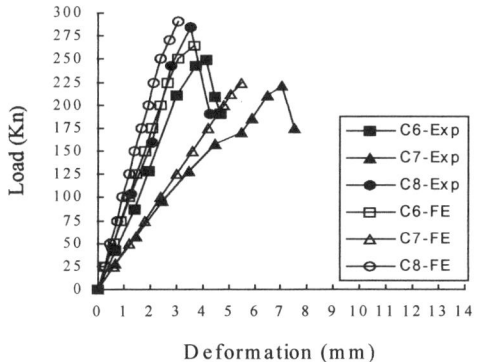

Figure 8. Comparison between Load Deformation Curves of Column Groups Pre-loaded to 67% of the Ultimate Load.

It can be seen from these results that the predicted values are in good agreement with the experimental ones.

The results obtained by the finite element analysis showed identical trend to those obtained experimentally. Comparing the analytical results predicted after the ferrocement layer has been activated to those of the original model, better load carrying capacity, ductility ratio, and energy absorption were obtained. For column group C1, the increase in the load carrying capacity, ductility ratio, and energy absorption were found to be 41.9%, 17.65%, and 97.65% analytically whilst the results measured experimentally were 46.8%, 26.6%, and 92.7%; respectively.

It was noticed from figures (5), (6), and (7) that the predicted load-deformation curves showed relatively higher stiffness than the experiment ones and the predicted ultimate loads were slightly higher than those measured experimentally. The difference between the analytical and experimental results in the ultimate loads ranged from 1.7% to 7.5%. However, higher difference in the deformation at ultimate loads was observed. It is well known that finite element models give higher stress predictions yet lower deformations than those measured experimentally. Due to the limitations of the educational version of ANSYS and the limited storage capacity, the behavior of the repaired column models could not be predicted beyond the ultimate load though a displacement control loading was followed. Therefore, the part of the experimental load-deformation curve beyond the ultimate load was neglected when comparing energy absorption.

From the comparison given in Table V, it is clear that finite element models slightly overestimate the ultimate loads yet they underestimate the energy absorption. However, the increase in P_u is not greater than 7.5% while the reduction in energy absorption is not greater than 16%. The difference in ductility ratio ranged from 8% decrease to 15% increase. This difference may be attributed to the difference in deformation between predicted and measured values.

CONCLUSION

Based on the results and observations of the experimental and analytical investigations presented in this paper the following conclusions can be drawn:

- Under concentric loading conditions, reinforced concrete columns damaged due to over-loading can be restored with enhanced strength and performance using ferrocement jackets.
- After repairing, all test specimens and analytical models showed higher deformation at ultimate load, increase in the ductility ratio, and considerable increase in energy absorption.
- High pre-loading levels resulted in lower gain in the load carrying capacity and energy absorption, yet increase the gain in the ductility ratio. Columns pre-loaded to working load (67%) showed higher gain in the ultimate load and energy absorption than those pre-loaded to higher levels; however, the increase in the ductility ratio was the least.
- The steel ratio of the ferrocement jacket has a great influence on the gain percentage of the ultimate load, ductility ratio, and energy absorption. The higher the volume fraction the higher the gain in P_u and the lower the gain in ductility ratio and energy absorption.

- For the same volume fraction, the welded wire mesh exhibited higher gain in the load carrying capacity than the expanded metal lath. However, the expanded metal showed higher gain in ductility ratio and energy absorption.

REFERENCES

1. Mansur M.N., and Paramasivam P. 1990. "Ferrocement short columns under axial and eccentric compression," ACI structural Journal, Vol.87, No.5, pp. 523- 529.
2. Kaushik S.K., Prakash A., and Singh K.K. 1994. "Inelastic buckling of ferrocement encased columns," Proceedings of the fifth international symposium on Ferrocement. Edited by Nedwell and Swamy, E& FN Spon, London, pp. 327- 341.
3. Ahmed T., Ali SK. S., Choudhury J. R. 1994. "Experimental study of ferrocement as a retrofit material for masonry columns," Proceedings of the fifth international symposium on Ferrocement. Edited by Nedwell and Swamy, E& FN Spon, London, pp. 269- 276.
4. Nedwell P.J., Ramesht M.H., and Rafei-Taghanaki S. 1994. "Investigation into the repair of short square columns using Ferrocement," Proceedings of the fifth international symposium on Ferrocement. Edited by Nedwell and Swamy, E& FN Spon, London, pp. 277- 285.
5. Fahmy, E., and Korany, Y. 1996. "Application of Laminated Ferrocement for Repairing Reinforced Concrete Slabs," The Annual Conference of The Canadian Society for Civil Engineering, Edmonton, Alberta, Canada, pp. 14- 25.
6. Fahmy, E., Shaheen, Y., and Korany, Y. 1997. "Repairing Reinforced Concrete Beams by Ferrocement," Journal of Ferrocement, Bangkok, Thailand, Vol.27, pp. 19- 32.
7. Korany, Y.1996."Application of Laminated Ferrocement in Repairing R.C. Slabs and Beams," M.Eng. Thesis, The American University in Cairo, Egypt, pp. 70-72.
8. Swanson Analysis Systems Inc. 1997. "Theoretical Manual," ANSYS Engineering Analysis System, Version 5.4, Huntington Beach, Calif., USA.
9. Paramasivam, P., Nathan, G. K.1984. "Prefabricated ferrocement water tanks," ACI Journal, pp.580-585.

Earthquake Rehabilitation of a Historic Concrete Structure Using Fluid Viscous Dampers

H. K. MIYAMOTO and D. A. LEE

ABSTRACT

Hotel Woodland is one of the first structures in North America to be seismically retrofitted using viscous dampers (VDs). VDs were selected because it was essential to maintain the historical appearance of this 4-story 1927 vintage Historical Landmark reinforced concrete building located in Woodland, California at a competitive cost.

The building is essentially a non-ductile reinforced concrete (RC) frame at the first level, and RC shear wall at levels 2, 3, and 4. Retrofit options included adding conventional shear walls or braces at the first level, and using VDs and steel moment frames at the first level.

Adding shear walls or braces at the first level would have seriously altered the historical appearance of the building and also would have limited commercial development. Computer Analyses revealed that installing VDs and moment frames at the first level reduced drifts at all levels to the desired performance. Using VDs proved to be the most cost effective method for seismically retrofitting the building and preserving the historical appearance of the building.

This paper presents a case study for retrofit criteria that considers earthquake demand, building response performance, historical interests and economic consideration.

H. Kit Miyamoto, S.E., Marr-Shaffer & Miyamoto, Inc., 1661 Garden Highway, Suite 101, Sacramento, CA 95833 (916) 567-0793
David A. Lee, Taylor Devices, 2118 Wilshire Boulevard, Suite 604, Santa Monica, CA 90403 (310) 396-5459

INTRODUCTION

Seismic rehabilitation of existing buildings is one of the most challenging tasks that structural engineers face today. Historical buildings are particularly difficult because of their strict architectural requirements. Every building is unique in its own way, and there is no easy cookbook approach. Conventional codes and methods may not be applicable, so engineers must be creative as they tackle retrofitting problems. Adequate communication between the design team and the owner of a building is critical, much more so than with projects involving new buildings. Performance objectives, cost limitations and future commercial development are very important issues that everybody must understand and agree upon.

The Northridge Earthquake of 1994, put into question, the fundamental concept of earthquake resistant design, which relies heavily on the ductile behavior of the material. Code-confirmed concrete and steel buildings were damaged beyond engineers' expectations. Supplemental damping is one possible solution to this problem. Dampers can eliminate or reduce plastic deformation of members, which decreases the uncertainty involved with nonlinear behavior of the structure.

This paper presents results of a case study illustrating the processes and decisions regarding the seismic retrofit design of a four-story non-ductile reinforced concrete building.

DESCRIPTION OF THE STRUCTURE

Hotel Woodland is a 4-story reinforced concrete building constructed during the latter part of 1927. The building is a National Historic Registered building. The ground level footprint is approximately 168 feet by 95 feet, the upper three levels have a footprint of 168 feet by 50 feet, and the total square footage is approximately 50,000 ft^2. The total height of the structure is about 53 feet, not including the basement which is under only part of the building. The ground floor is used as commercial/retail space and the 2nd floor and above are single-occupancy apartments presently occupied (See Figure 1).

None of the original plans were available at the time of analysis, therefore destructive investigation was conducted to determine material properties. The 2nd, 3rd, and 4th floors are cast-in-place concrete joist-beam construction with 2 1/2" concrete slab. Typical columns are 16" square concrete, reinforced with 4-3/4" square grade 40 reinforcing bars with 5/16" square ties at 12". Typical exterior frame consists of 48" deep by 10 3/4" thick concrete spandrel beams and 48" wide by 6 3/4" thick concrete piers. The concrete wall pier-spandrel beam construction is terminated at the 2nd floor. In addition, 6 3/4" thick concrete bearing-shear walls exist at the East and West end of the structure at the ground floor.

No lateral resisting elements were found at the North and South elevation of the building at the ground floor, except for 16" square lightly reinforced concrete columns. This type of structure in the East-West direction is often defined as a non-ductile soft/weak story structure. The total reactive weight of the structure is approximately 5100 kip and the destructive testing indicated that the average compression strength of the concrete is approximately 3000 psi.

Figure 1. Architectural Elevations

The subsurface investigation revealed that the structure is supported on isolated spread foundations extending five feet below the slab on grade. The typical dimensions are 11 feet by 11 feet by 1 foot deep. The geotechnical consultant determined that a site coefficient of S_2 was appropriate for the site.

SEISMIC RETROFIT CRITERIA

This seismic retrofit was at the option of the owner, so maintaining the historical appearance of the building, improving the earthquake performance of the building, and minimizing the cost were the primary considerations in establishing the retrofit design.

When seismically rehabilitating existing structures, use of the Uniform Building Code is insufficient since it does not address performance objectives other than life safety in quantifiable ways (Hart & Elhassan, 1994 [1]). Therefore, the design team and owner defined the design seismic event and the performance criteria for the structure.

Performance criteria is based on the Uniform Code for Building Conservation, which defines a performance objective as "promote public safety and welfare by reducing the risk of death or injury that may result from the effects of earthquakes" (ICBO, 1994 [2]). Considering the cost, the design team and owner decided that the seismic retrofit objective should be limited to preventing the collapse of the four-story super structure, since it would present a major threat to life safety. Some damage to the super structure was allowed.

The Design Basis Earthquake (DBE) is a 20% probability of occurrence in a 50-year duration. This event is consistent with the California Seismic Safety Commission Recommendations for the "Acceptable Seismic Risk for State Buildings" report. The Maximum Capable Earthquake (MCE) selected for the retrofit is a 10% probability of occurrence in a 100-year duration. Three pairs of Time Histories for each DBE and MCE event were constructed to analyze the structure. A detailed discussion regarding the design usage of DBE and MCE is described in the 'Design Criteria' section of this paper.

SEISMIC HAZARD

The geotechnical consultant (Wallace-Kuhl & Associates) performed a site-specific ground motion study for the hotel site. Historically, the largest earthquake event to influence the site was the Richter Magnitude 6.75 Vacaville-Winters event of 1892. This event and its aftershocks were estimated to have produced an attenuated site acceleration of approximately 0.14 g (Gius, 1994 [3]).

Deterministic and probabilistic analyses were performed to estimate the Peak Ground Acceleration (PGA). These analyses revealed that 0.17g site acceleration would occur from a magnitude 4.5 DBE event on the Dunnigan Hills fault and .26g acceleration would occur from a magnitude 5.5 MCE event on the Dunnigan Hills fault [3].

Actual California earthquake time histories were utilized to develop site-specific ground response spectra. Horizontal ground accelerations were selected based on events 15 to 30 miles from the recording station, at an alluvium underlain station, and from an event of a similar fault mechanism to faults expected to affect the site (thrust faults). The selected horizontal ground time histories were scaled to DBE and MCE maximum horizontal accelerations considering potential amplification effects of the soil by a computer program 'Shake' [3], (See Figure 2).

EXPECTED PERFORMANCE OF THE EXISTING BUILDING

As part of the study, 3-dimensional Time History analysis of the original building was performed. An analytical model was subjected to non-reduced DBE Time Histories. Due to the lack of ductility details, hysteric energy absorption capacity of the existing material was discounted. Following are the results of the study:

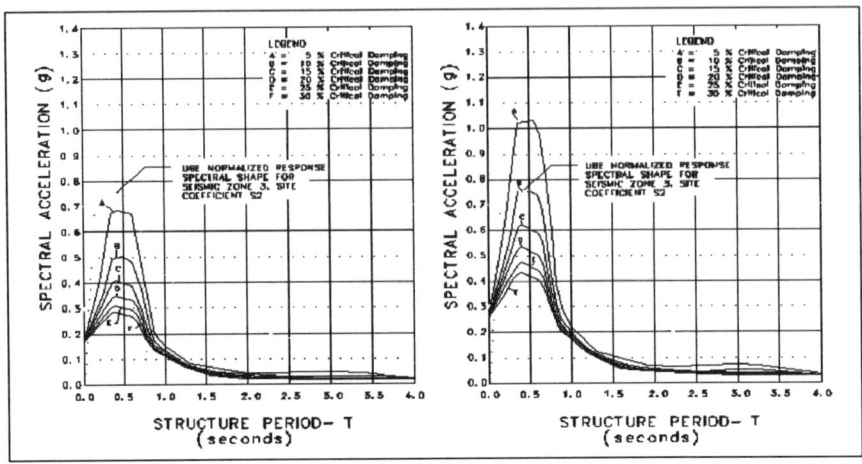

Figure 2. Site Specific DBE Spectra, and Site Specific MCE Spectra

East-West (Longitudinal) Direction:

The fundamental period of the building in the East-West direction was approximately .7 second. Effective mass factor of the fundamental mode was 98%. Maximum displacement and story shear are shown in Table 1.

TABLE (1)

Story	Story Displacement (inch)	Story Shear (kips)
Roof	0.01	290
4th	0.02	890
3rd	0.01	1484
2nd	2.00	2200 (.43G)

The concrete columns at the ground floor level were overstressed in bending and shear due to excessive deflection and the lack of ductility detailing and strength. Most of the non-linear behavior of this building was concentrated at the ground floor level columns. This type of adverse behavior could cause total collapse of the superstructure. An example can be seen at Olive View Hospital after the 1971 San Fernando earthquake. Under earthquake-induced loading, excessive lateral displacement caused plastic hinges to form in the ground floor columns (Moehle, 1994 [4]).

North-South (Transverse) Direction:
Fundamental period of the building in the North-South direction was approximately 0.16 second. Effective mass factor of the fundamental mode was 84.1%.

The wall piers and shear walls at the East and West ends of the building were overstressed in shear. This type of failure could cause total collapse of the structure. Also, torsional excitation of the building was excessive due to the substantially longer shear wall at the West end.

SEISMIC RETROFIT SCHEMES

After the design team and the owner defined and agreed on the above retrofit criteria, numerous seismic retrofit schemes were considered. Since this building is a National Historical Registered building, there were unique challenges that the design team had to meet including: 1) keeping the historical appearance of the landmark hotel, 2) maximizing the retail/commercial area at the ground floor, 3) avoiding disturbance to tenants living in apartments on the 2nd floor and above, and 4) managing cost effectiveness.

East-West (Longitudinal) Direction:
Concrete shear walls were rejected because they caused two ancillary problems: I) damage of historical appearance, and 2) limitation of commercial development. Conventional steel brace frames were considered and rejected because practical brace locations were very limited. Also the steel brace frame and shear wall construction required major foundation modification. Concrete jacketing of existing ground floor columns was considered and rejected because of the uncertainty of the existing construction and cost limitation. Finally, using steel moment frames with fluid viscous dampers (VDs) at the ground floor was chosen.

The steel moment frames were designed to provide stiffness, strength, and redundancy, which the existing lightly reinforced concrete columns lacked. VDs were provided to control drift at the 1st floor and to keep steel moment frames in the elastic range. Elastic rotation of the moment frame connection was limited to acceptable level. VDs were attached to the top of the steel Chevron Braces (See figure 3) and VDs-Chevron Braces were strategically located to meet the above requirements.

Fluid viscous dampers were selected over other supplemental dampers for the following reasons: 1) since it is a velocity-dependent system, large energy dissipation would be activated with small displacement, 2) the forces in VDs are out of phase with axial loading of the columns, and 3) the long history of military application proves system reliability.

North-South (Transverse) Direction:

Displacement and velocity of the existing concrete shear walls were not large enough to activate VDs, therefore conventional shotcrete and new shear walls were provided at the ground floor level. The shotcrete and the new shear wall did not damage the historical appearance of the structure in the North-South direction.

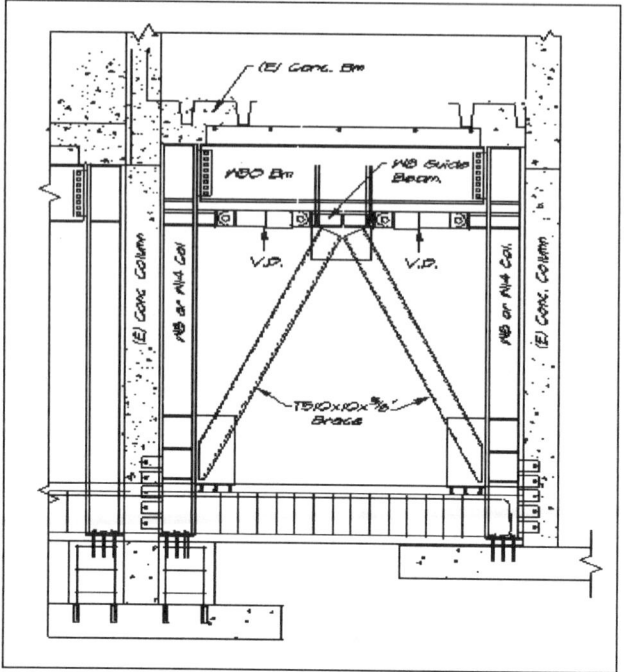

Figure 3: Type VD Assembly Elevation

DESCRIPTION OF THE FLUID VISCOUS DAMPERS

Fluid viscous dampers (VDs) use the flow fluid through an orifice to provide a resistive force. A piston travels through a chamber that is filled with silicone oil. The pressure difference across the piston head creates the damper force. Seismic energy is transformed into heat, which dissipates into the atmosphere (See figure 4). VDs can operate over temperatures from -40 F to 160 F. The orifice construction utilized is similar to that in the classified application for the U.S. Air Force B-2 Stealth Bomber and is considered state-of-the art (Constantinou & Symans, 1992 [5]).

Sixty-six earthquake simulation tests were performed on 1 and 3-story model structures at the State University of New York, Buffalo by Constantinou and Symans. The damper exhibited essentially linear viscous behavior for a range of frequencies below 4 Hz. The addition of VDs in the tested structure resulted in 30-70% reduction in story drifts and 40-70% reduction in story shear forces.

ANALYTICAL PROCEDURE

Two different mathematical models of the building were constructed to study retrofit schemes. One was a simple 2 dimensional stick model and the other was a complex 3-dimensional finite element model.

The 3-dimensional model was used to verify the 2-D stick model results, predict member forces, and study cross coupling between stiffness of braces and VDs. Time history analyses were performed using the computer program ETABS 6.04 (CSI, 1994 [6]) which also utilized the Step-by-Step Linear Acceleration Method. The computer model had 146 column lines, 115 beam bays, 16 brace elements, 22 panel elements, and 8 link elements.

DESIGN CRITERIA

Three Time Histories for each DBE and MCE event were chosen and scaled based on the criteria described in the 'Seismic Criteria' section of this paper. These time histories include:

1) Coalinga 1983 Cantua Creek School Channel 1
2) Coalinga, 1983, Channel 3, and
3) El Centro 1940 A-S Record

Each event was applied simultaneously to the mathematical model in X and Y directions. The analysis indicated that Coalinga 1983, channel I, produced the worst case scenario, therefore it was chosen as the design earthquake motion.

Critical elements in the structure were designed to sustain limited damage for the MCE event, and the other elements were designed for the DBE event. The following is a summary of design criteria:

1) Critical ground floor concrete columns were analyzed with the MCE event, considering cracked sections, and p-delta effect.
2) The ground story drift was limited to 0.002 at DBE and 0.003 at MCE to protect the existing brittle structure.
3) All existing and new shear wall responses were limited to elastic range only at the DBE event.
4) The foundation stability was analyzed using the DBE event.
5) The stress ratio of the new Steel Moment Frames was limited to approximately 20% of the yield to protect the welded connections at the DBE event.
6) VDs were designed for MCE events. All VD connections and Chevron Braces assembly were also designed to remain elastic for MCE events.

EXPECTED PERFORMANCE OF THE RETROFITTED STRUCTURE

East-West (Longitudinal) Direction:

The fundamental period of the retrofitted building in the East-West Direction was .46 second. The effective mass factor of the fundamental mode in the East-West Direction was 97%. Approximately 40% of critical damping was provided by VDs at the ground floor where maximum inter-story drift and velocity occurred. A total of 16 steel moment frames with W 14x132 grade 50 columns, and W 30x99 grade 50 beams were provided. A total of 8-VD assemblies with *16-50* kip output dampers were provided (See figure 4). The damping constant for each VD was 9.4 kip-second/inch. The exponential constant was set as a unit, which produced perfect linear viscous behavior. The maximum design axial force of the VDs was 100 kip with a safety factor of 2.0. The maximum displacement, velocity, and story shear for DBE are shown on Table 2 for 5% of critical damping without VDs and on Table 3 for 40% of critical damping at the ground level with VDs.

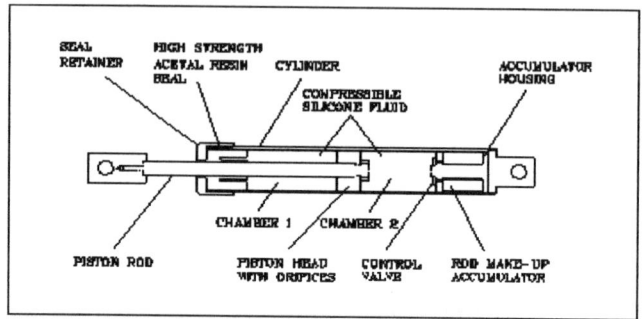

Figure 4. Construction of Fluid Viscous Damper

TABLE (2): STEEL MOMENT FRAME WITHOUT VDS (DBE), 5% MODAL DAMPING

Story	Story Displacement (inch)	Story Shear (kips)	Story Velocity (inch/sec)
Roof	0.01	2090	0.41
4th	0.03	1256	0.25
3rd	0.03	2090	0.41
2nd	1.31	3087 (.60G)	16.80

TABLE (3): STEEL MOMENT FRAMES WITH VDS (DBE): FINAL DESIGN, 40% MODAL DAMPING.

Story	Story Displacement (inch)	Story Shear (kips)	Story Velocity (inch/sec)
Roof	0.01	183	0.05
4th	0.01	558	0.13
3rd	0.01	927	0.22
2nd	.41	1374(.26G)	5.50

The above tables show that by providing VDs, both base shear and 2nd floor displacement were reduced by approximately 60%. Plastic deformation of both existing concrete and new steel moment frames were precluded, and the majority of the seismic energy was absorbed by VDs.

North-South (Transverse) Direction:

Fundamental period of the retrofitted building in the North-South direction was 0.16 second. The effective mass factor of the fundamental mode in the N-S direction was 82%. Five-inch shotcrete was added to the existing 7" wall at the full height of the East-West elevation. Also a new 12" shear wall was added at the ground floor to reduce torsional excitability of the structure.

REHABILITATION COST

Total construction cost of the seismic strengthening was approximately $500,000 U.S. (1995 present) which equates to $10.00 per square foot. The above figure satisfied the construction cost requirement of the project. Construction began in August 1995.

CONCLUSION

A combination of steel moment frames and VDs proved to be the most cost-effective method for seismically retrofitting the building. It also accommodated the historical appearance and commercial utilization requirements. In addition, limiting plastic deformation of the structural material reduced the uncertainty in the structural behavior in the case of a seismic event.

Using supplemental damping can be a very effective method to resist seismic force for many buildings. The authors strongly believe that supplemental dampers will be one of the 'star' solutions to protect structures from the destructive forces of earthquakes in the 21st century.

REFERENCES

1. Hart, G. and Elhassan, R. (1994). "Seismic Strengthening with Minimum Occupant Disruption Using Performance-Based Seismic Design Criteria." Earthquake Spectra, Earthquake Engineering Research Institute, Volume 10, Number 1.
2. ICBO (1994). "The Uniform Code for Building Conservation". International Council of Building Officials, Whittier, CA.
3. Gius, D. (1994). "Geotechnical Engineering and Seismic Design Parameters". Wallace-Kuhl & Associates, Sacramento, CA.
4. Moehle, J. (1994). "Design and Detailing of Moderately Tall Wall Buildings". Proceedings of Advances in Earthquake Engineering Practice. University of California, Berkeley, Berkeley, CA.
5. Constantinou, M. and Symans, M. (1992). "Experimental and Analytical Investigation of Seismic Response of Structures with Supplemental Fluid Viscous Dampers", NCEER-92-0032, State
University of New York at Buffalo, Buffalo, New York.
6. CSI (1994). "ETABS version 6.04.", Berkeley, Ca.

Seismic Connection Designs for Retrofitting Steel Moment Frames

J. ALLEN and R. M. RICHARD

Since the January 17, 1994 Northridge, California Earthquake, many fractured steel moment frame connections have been found and repaired. The cost of repair for a fractured bottom or top beam flange connection is of the order of $6,000 to $20,000 depending upon the mode of failure, type of building skin, connection location within the building, type of repair, etc.. Detailed analytical studies of typical wide flange beam to column connections to determine stress distribution at the beam/column interface had not been made prior to the studies presented herein. Strain rate concentrations, rise time of applied loads, stress concentration factors, stress gradients, residual stresses and geometrical details of the connection all contribute to the behavior and strength of these connections. By using high fidelity finite element models and analyses to design full scale experiments of a test specimen, excellent correlation has been established between the analytical and test results of stress and strain profiles at the beam/column interface where fractures occurred. Location of the strain gauges on the beam flange at the column face was achieved by proper weld surface preparation which confirmed the analytically determined high strain gradients and stress concentration factors. Stress concentration factors between 4.5 and 5.0 were found at the root center of the beam flanges for a typical W27x94 beam to W14x176 column pre-Northridge connection with no continuity plates.

A new design detail using a slotted beam web by Seismic Structural Design Associates, Inc. (SSDA) which reduces the high non-uniform stresses and strain rates that exist with conventional designs has been analyzed and tested. The slotted beam web designs reduce the Stress Concentration Factor (SCF) at the beam-to-column flange connection from a typical value of 4.6 down to a typical value of 1.4 by providing a near uniform flange/weld stress and strain distribution. This 4.6 SCF, computed by finite element analyses and observed experimentally, exists in the pre-

---

J. Allen, The Allen Company and Seismic Structural Design Associates, Inc. 30131 Town Center Dr. #220, Laguna Niguel, CA 92677
R. Richard, Professor Emeritus, University of Arizona, and President of Seismic Structural Design Associates, Inc., Tucson, Arizona

Northridge, reduced beam section (dogbone), and cover plate connection designs and results from a large stress and strain gradient <u>across</u> and <u>through</u> the beam flange/weld at the face of the column. For ductile materials this reduction in the SCF results in a reduction of the ductility demand in the material at the column flange/beam flange/weld by about an order of magnitude.

The slotted beam design (1) develops the full plastic moment capacity of the beam, (2) moves the plastic hinge in the beam away from the face of the column, and (3) results in near uniform tension and compression stresses and strains <u>across</u> and <u>through</u> the beam flanges from the face of the column to the end of the slot, (4) dramatically reduces the beam flange/weld vertical shear forces and (5) reduces the connection residual weld stresses. Moreover, the slotted beam design allows the beam flanges to buckle independently from the beam web so that the amplitude of the lateral-torsional plastic buckling mode that occurs in the non-slotted connections is very significantly reduced. This latter attribute reduces the torsional moment and torsional stresses in the beam flanges and welds at the column flange.

SSDA has performed extensive FE analyses that demonstrate the flange/weld stress distribution is dramatically improved in the slotted beam connection by reducing the flange "prying" moment at the face of the column. For example, the stress and force distributions for a W36x150 beam connected to a W14x311 column in an ATC-24 test protocol mode. This fully welded assembly was tested (dynamically) at Lehigh University and partial results were reported in Modern Steel Construction (MSC), January 1996. Continuity plates 1" thick that match the 0.94" beam flanges are provided. SSDA made three FE elastic analyses of this assembly: baseline (pre-Northridge), dogbone (DB2 specimen with a radius cut - MSC August 1996), and the SSDA Slotted Web.

These FE analyses showed that the tension flange/weld stress distribution horizontally across and vertically through the beam flange is essentially identical for both the baseline and dogbone connections. Both have a large vertical stress gradient of 206 ksi at the center of the flange (whereas the slotted beam has a gradient of approximately 25 ksi). Note that the middle half of the bottom surface of the tension flange/weld of the nonslotted beam is in compression. If a backup bar were in place on this tension flange, half of it would also be in compression. However, if the loading is reversed so that the bottom flange is in tension, then the backup bar would have a tensile stress of 176 ksi at its center. It seems very apparent that this, along with the large strain gradients that require a large ductility demand, explains the predominance of bottom flange/weld and column divot fractures.

These FE analyses also showed that both the baseline and dogbone connection designs resulted in large vertical shear forces in the beam flanges and the beam-to-column welds (50% of the total vertical shear). After performing analyses similar to these, Subhash Goel at University of Michigan recommends that a combined cover plate/vertical beam flange fins design be made to resist all the beam vertical shear at the column (he presented this at the Japan-SAC workshop of Steel Building in Earthquakes, February 1997). However, the slotted web design essentially eliminates the vertical shear in the beam flange weld in addition to very substantially reducing the stress and strain gradients and prying action in the beam flange/weld.

An existing ten (10) story welded steel moment frame (WSMF) structure located in the City of Burbank, California was retrofitted in the Spring, 1996 utilizing the SSDA connection. The Allen Company (TAC) was retained by HFB-Glenoaks Associates, LLC to (a) review the existing property located at 303 North Glenoaks Boulevard, Burbank, California, (b) review reports prepared by other consultants, (c) perform lateral analysis, (d) prepare construction drawings for the seismic retrofitting, (e) provide finite element analysis of two (2) representative assemblies, and (f) provide full scale testing of said assemblies. The existing property consists of a ten (10) story office tower with a basement and a five (5) level parking structure. The structures are located within twenty (20) miles from the epicenter of the 1994 Northridge Earthquake.

Following the Northridge Earthquake, the moment connections were tested by Twining Laboratories. On January 6, 1995, Twining Laboratories tested five (5) moment connections and found weld discontinuities in five (5) bottom flange connections and one (1) top flange connection. In March 1995, Twining Laboratories tested twenty-six (26) moment connections for John A. Martin & Associates, Consulting Structural Engineers, and found weld discontinuities in twenty-five (25) bottom flange connections and nine (9) top flange connections.

The office tower consists of a basement level below grade, and ten (10) stories above grade with a mezzanine level between the plaza level (ground level) and second floor level. The building dimensions above grade are 40.41m x 40.41m. Typical floor to floor height is 4.12m. The typical structural framing consists of 6.35cm hard rock concrete topping over 7.62cm - 20 GA metal decking supported by steel beams and columns. The lateral force resisting system above grade consists of four (4) welded steel moment frames (WSMF) at the building perimeter. Each side of the building has one WSMF. Each WSMF has three (3) 9.144m bays. The lateral force resisting system below grade consists of 30.48cm thick concrete shear walls at the perimeter of the building. The concrete walls are supported by continuous footings and the continuous steel columns are supported by spread footings.

TAC performed three dimensional dynamic analysis of the structure for the lateral forces of the present Code (1994 Uniform Building Code). The typical repair detail as shown in the construction drawings consists of (1) new 1.911cm thick stiffener plates at the bottom of beam bottom flange, (2) fillet welding the shear plate to the beam web, (3) square groove welding of the beam web to the column flange, (4) beam slots in the beam web near the top and bottom flanges. This retrofit detail is based on SSDA's full scale test results and the finite element analyses of beam-to-column connections.

SSDA conducted full-scale tests of two beam-column connections at Smith-Emery Company (SEC). Test Specimen 5A consists of a W33x130 beam and a W14x283 column and was tested following ATC-24 protocol on January 29 - February 1, 1996. The assembly was constructed by SEC as shown in the attached details prepared by TAC. The beam web is welded to the column flange. The beam slots are 1.5 X the beam flange width in length or approximately 45.72cm long. Two vertical fins occur at the top of the top beam flange and bottom of the bottom beam flange. A normal shear plate is utilized. Test SSDA-5A achieved 0.056 radians inelastic rotation. Test Specimen 7 consists of a W36x170 beam and a W14x283 column (which is typical beam-to-column connection in the existing office tower) and was tested in March 7, 1996. The beam slots are 1.5 X the beam flange width in length or approximately 45.72cm long. The shear plate is fillet welded to the column flange; however, the beam web receives a square groove weld to the column flange. This specimen achieved an inelastic rotation of 0.056 radians (See Figure 1).

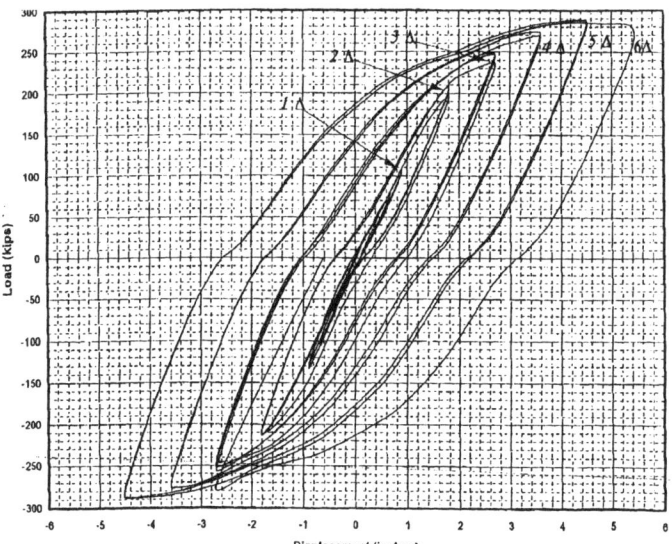

Figure (1): Load/Displacement Hystrises of Beam-Column Joint

The SSDA test results and the finite element analysis indicate that the beam slots are beneficial in reducing the ductility demand for the moment connection. Based on (a) the stress ratios from the dynamic analysis of the office tower and (b) the field test results of the one-hundred and fourteen (114) moment connections, TAC selected one-hundred and ninety-four (194) connections to be repaired out of two-hundred and sixty-four (264) total moment connections (approximately 73.5%). TAC did not select any connections to be repaired at the roof, 10th level, and at the plaza level since the seismic girder/column stress ratios are low. The stress ratios at the plaza level are small because the lateral force resistance is provided by the 30.48cm thick concrete shear walls.

REFERENCES

1. Allen, J., Partridge, J., Richard R., "Stress Distribution in Welded/Bolted Beam to Column Moment Connections", The Allen Company, March, 1995.
2. Richard, R., Partridge, J., Allen, J., and Radau, S., "Finite Element Analysis and Tests of beam-to-Column Connections", Modern Steel Construction, AISC, Volume 35, No. 10, p. 44-47, October, 1995.
3. Allen J., Partridge, J., Radau, S., and Richard, R., "Ductile Connection Designs for Welded Steel Moment Frames", Proceedings-Structural Engineers Association of California, 64th Annual Convention, October, 1995.
4. Richard., R., Allen, J., and Partridge, J., "Proprietary Slotted Beam Connection Designs", Modern Steel Construction, AISC, p. 28-33, March, 1997.
5. American Institute of Steel Construction, AISC Northridge Steel Update I, October, 1994.
6. Collins, J.A., "Failure of Materials in Mechanical Design (2nd ed.)", New York, John Wiley & Sons, 1993.
7. Youseff, N., Bonowitz, D., Gross, J.L., " A Survey of Steel Moment Resisting Frames Affected by the 1994 Northridge Earthquake", Rep. No. NISTIR 5625, NIST, Gaithersburg, MD.
8. Uniform Building Code" Volume No. 2, Structural Engineering Design provisions, pages 2-361, 1994.
9. "SEAOC Blue Book", Structural Engineers Association of California Seismic Design Code, 1996.
10. Englehardt, M.D., Sabol, T.A., and Frank, Karl., "Testing Connections", Modern Steel Construction, AISC, p 36, May, 1995.
11. Uang, C-M, Bondad, D., "Dynamic Testing of Pre-Northridge Steel Moment Connections", Rep. No. SSRP- 96/02, University of California, San Diego, CA, 1996.
12. Kaufmann, E., Ming Xue, Le-Wu Lu, Fisher, J., "Achieving Ductile Behavior of Moment Connections", Modern Steel Construction, AISC, p. 30, January, 1996.
13. Zekioglu, Atila, Mozaffarian, H., Le Chang, King., and Uang, C-M, "Designing After Northridge", Modern Steel Construction, AISC, p. 36, March, 1997.
14. Neuber, H., H., "Theory of Stress Concentration for Shear Strained Prismatic Elements with Nonlinear Stress- Strain Law", Jr. of Applied Mechanics, Vol. 28, Series E, No. 4, December, 1961.

Author Index

Abdalla, H., 73
Al-Chaar, N., 181
Alameddine, F., 46
Allen, J., 234

Chakrabarti, P., 59

do Valle, C., 73

Elhassan, R. M., 96
Elsanadedy, H. M., 59, 108, 119
Engebretson, D., 171

Fahmy, E. H., 210
Feng, M., 85
Frett, E., 147

GangaRao, H. V. S., 197
Gergely, J., 181

Haroun, M. A., 85, 108, 119
Hartley, A., 171
Hollaway, L, 26
Hooks, J. M., 1, 181
Horeczko, B. N., 187

Imbsen, R. A., 46
Islam, M. S., 73
Issa, M. A., 73

Kachlakev, D., 131
Korany, Y. S., 210
Kreiner, J., 119, 162

Lancey, T., 119, 162
Lansburg, S., 159
Lee, D. A., 224
Leslie, M., 73

McCurry, D. D., Jr., 131
Miyamoto, H. K., 224
Mosallam, A., 59, 85, 119, 162
Mullins, G., 171

Richard, R. M., 234
Roberts, J., 11

Sen, R., 171
Shaheen, Y. B. I., 210
Shekar, V., 197
Sim, S., 59
Stevens, G. R., 141

Thippeswamy, H., 197

Young, D. T., 181
Youssef, M., 85